FLOWS, BOUNDARIES, INTERACTIONS

FLOWS, BOUNDARIES, INTERACTIONS

Sinaia, Romania 3 – 5 May 2007

EDITORS

Cristiana Dumitrache
Vasile Mioc
Nedelia A. Popescu

Astronomical Institute of Romanian Academy
Bucharest, Romania

SPONSORING ORGANIZATION

The Research and Education Ministry
under the Program Excellence in Research

Melville, New York, 2007

AIP CONFERENCE PROCEEDINGS ■ 934

Editors

Cristiana Dumitrache
Vasile Mioc
Nedelia A. Popescu

Astronomical Institute of Romanian Academy
Str. Cutitul de Argint 5
040557 Bucharest
Romania

E-mail: crisd@aira.astro.ro
vmioc@aira.astro.ro

L.C. Catalog Card No. 2007934166
ISBN 978-0-7354-0445-8
ISSN 0094-243X

Printed in the United States of America

CONTENTS

SUN AND HELIOSPHERE

STARS AND GALAXIES

PREFACE

In 2007 we celebrated fifty years from the International Geophysical Year and extended our scientific horizon to that of Heliosphere. During these fifty years of space era, our knowledge about the Sun and stars, galaxies or our Galaxy, advanced very much indeed. Now, in the International Heliophysical Year, we focus on few scientific themes that we intended to approach in this workshop, themes as:

- Generation and evolution of magnetic structures in solar atmosphere and Heliosphere;
- Transient phenomena;
- Flows and circulations;
- Boundaries and interfaces;
- Sun and Heliosphere in 3D;
- Solar stars;
- Dynamics of the Galaxy and related problems.

The workshop was divided in two important sections:

(1) Sun and Heliosphere;

(2) Stars and galaxies.

The first section treated solar magnetic structures such as active regions, flares and CMEs, prominences and coronal streamers. Both observational and models works were presented. Flows, boundaries and interfaces were approached in numerical simulations. Interface between streamers and coronal holes and also solar corona rotation were deduced from SOHO data. Coronal waves and Sun quakes were also treated in few papers. Some papers came from dynamical systems or differential equations formalism to solar physics or heliosphere phenomena study.

This section had as invited B. Schmieder, T. Forbes, K. Petrovay, S. Poedts, J.L.Ballester, I. Ballai, G. Noci and S. Giordano.

The second section was devoted to stellar studies, extragalactic astronomy and especially cosmology, having as invited M.D. Suran.

This workshop constituted an opportunity to bring together personalities and young PhD students, to change ideas and results, to put the bases for future fruitful cooperations.

THE EDITORS

SUN AND HELIOSPHERE

The origin of magnetic helicity in solar active regions

K. Petrovay

Eötvös University, Department of Astronomy, Budapest, Pf. 32, H-1518 Hungary
and
Theoretical Institute for Advanced Research in Astrophysics, Dept. of Physics, National Tsing Hua University, Hsinchu 30013, Taiwan

Abstract. The magnetic fields in solar active regions are known to be helical. The current helicity, as measured from magnetograms, is negative on the northern and positive in the southern hemisphere, its normalized mean value being $\alpha_p \equiv \mathbf{B} \cdot (\nabla \times \mathbf{B})/B^2 \sim 10^{-8}\,\mathrm{m}^{-1}$. Recent observations have convincingly demonstrated that the helicity originates in subsurface regions and the flux tubes emerge in an already helical form. One possible contribution to this helicity is the accretion of the weak general poloidal magnetic field of the Sun by the rising toroidal flux loop. For this process to work, the field lines of the background field first need to be wrapped around the rising flux tube, which assumes that freezing-in holds to a good degree of approximation. The wrapped-up field then has to diffuse into the flux tube, giving rise to a twisted magnetic field structure. The conflict between the requirements of freezing-in (for wrapup) and diffusion (for penetration into the tube) is only apparent. The role of diffusion is likely to be limited during most of the rise of the tube, allowing a significant amount of flux to be swept up; then, in the uppermost 20–30 % of the tube's path through the convective zone diffusion prevails, allowing penetration of the wrapped up flux.

Keywords: magnetohydrodynamics, Sun, magnetic helicity
PACS: 47.65.-d, 95.30.Qd, 96.60.Hv

1. INTRODUCTION

The helical nature of magnetic structures in the Sun is important for several reasons. On the one hand, helicity is necessarily accompanied by magnetic free energy (i.e. excess magnetic energy compared to the potential field obeying the same boundary conditions). This excess energy in turn, can be released in fast reconnection processes, leading to fares and other eruptive phenomena. On the other hand, the helical nature of the field lends a certain coherence to magnetic configurations (such as prominences or interplanetary clouds) that might otherwise not even persist.

The helicity of magnetic structures can be observationally characterized by the *current helicity*, defined as

$$h_C = \mathbf{B} \cdot (\nabla \times \mathbf{B}) \tag{1}$$

where **B** is the magnetic field strength. For theoretical calculations the *magnetic helicity* is often preferred:

$$h_M = \mathbf{A} \cdot \mathbf{B} \equiv \mathbf{A} \cdot (\nabla \times \mathbf{A}) \tag{2}$$

where **A** is the magnetic vector potential. (To be precise, the above defined h_C and h_M are helicity densities, helicity being their integral over a finite volume.)

CP934, *Flows, Boundaries, Interactions*
edited by C. Dumitrache, V. Mioc, and N. A. Popescu

As Figure 1 illustrates, in a single flux rope helicity can take the form of either a twist or a *writhe* or *kink*.

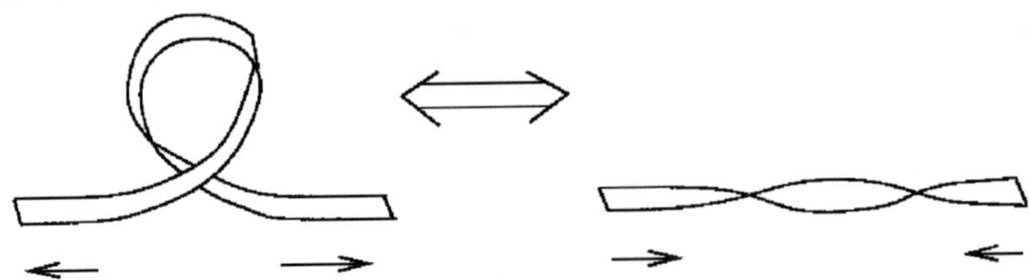

FIGURE 1. Twist and kink of a magnetic flux tube can be converted into each other, as easily demonstrated by pulling the ends of a kinked paper tape, then pushing them together again.

In the next section I briefly review the available observational evidence for magnetic helicity in solar active regions. Then I will describe in more detail one promising possibility to explain the origin of twist in the magnetic flux loops giving rise to active regions. This possibility consists in an accretion of poloidal flux by the rising toroidal flux tube giving rise to the active region, the source of this poloidal flux being the weak large-scale background magnetic field of the Sun. Section 4 discusses some conceptual difficulties related to this mechanism.

2. SOLAR OBSERVATIONS OF HELICAL MAGNETIC FIELDS

On images of the solar surface, especially on those taken in $H\alpha$, vortex-like, twisted or helical structures are often seen. Early claims by Hale [1] and Richardson [2] regarding the predominance of one chirality or another on a given hemisphere were hard to confirm owing to projection effects, a lack of information on 3D structure and large statistical scatter. Yet the existence of a hemispheric rule of chirality was convincingly demonstrated at least in the case of quiescent prominences and interplanetary magnetic clouds [3]. However, convincing direct evidence for chirality rules in active regions could only be provided by direct magnetic measurements.

Determinations of current helicity typically rely on a constant-α force-free magnetic field model, constructed using the measured line-of-sight field component as lower boundary condition. Such models are based on the assumption that $\nabla \times \mathbf{B} = \alpha_p \mathbf{B}$ where α_p is constant. Multiplying this relation by $\mathbf{B}$, it is clear that $\alpha_p = h_C/B^2$ is a normalized measure of current helicity, with a dimension of 1/length. Its meaning is easy to visualize, as $1/\alpha_p$ is the length along a flux tube over which he field lines make a full circle around the tube.

Of the possible values of α_p, a best fit is then chosen by comparing the resulting magnetic field structure to either morphological features seen in solar images taken in $H\alpha$, EUV, etc. [4] or, more recently, to the horizontal field components as measured by vector magnetographs [5]. A third method consists in simply taking the ratio $\alpha_p = (\nabla \times \mathbf{B})_n/B_n$, as measured by vector magnetographs, and averaging it over the active region. Results derived by different methods are generally in satisfactory agreement [6], [7].

All these methods currently yield a single mean value of α_p over the whole active region. This is because the low S/N ratio of vector magnetograph measurements inhibits a reliable study of current helicity distribution across the plage; for the same reason, usually only pixels with relatively high field strength ($B > 100\,\mathrm{G}$ or so) are used in the determination. It is expected that future improvements in magnetograph sensibility will remedy this situation.

It has now been firmly established by these studies that active region magnetic fields have helical structures, with a higher occurrence of negative helicity in the northern hemisphere. Observations show that the typical average value of the current helicity parameter α_p in an active region is on the order of $10^{-8}\,\mathrm{m}^{-1}$ [8].

The axes of bipolar active regions are tilted relative to the E-W direction. This is generally interpreted as evidence for writhe in the emerging flux tubes giving rise to active regions. The origin of this writhe is well understood: it is the consequence of the Coriolis force acting on the downflows in the flux loop legs [9]. The sign of current helicity implied by the observed tilts is positive on the northern hemisphere, i.e. opposite to the twist measured in the solar photosphere; its amplitude, however, remains well below the α_p values quoted above.

A basic question regarding the origin of the observed magnetic helicities is whether they are generated after the emergence of the flux loop or the flux emerges already in a helical form. Shearing of the photospheric footpoints of magnetic loops due to differential rotation or smaller scale granular and supergranular motions can, in principle, generate a helicity of the correct sign [10]; but the amount of helicity generated in this way seems to be insufficient to compensate for helicity losses in CME's [8]. In addition, in an important paper Pevtsov et al. [11] studied the time development of the helicity in young, emerging active regions and found a good correlation between emergence rate and helicity increase. In particular, further increase of helicity ceases once the emergence of the loop, as measured by the increase in footpoint separation and area growth, comes to a halt. This observation apparently settles the issue in favor of a subsurface origin of the observed twist.

3. HELICITY GENERATION BY THE ACCRETION OF POLOIDAL MAGNETIC FIELDS BY AN EMERGING FLUX TUBE

A possible theoretical explanation of the observed helicity was proposed by Choudhuri [12]. He suggested that the poloidal flux in the solar convection zone (SCZ) gets wrapped around a rising flux tube, as sketched in Fig. 2. Subsequently, Choudhuri et al. [13] showed that this mechanism can give rise to a helicity of the same order as what is observed. Choudhuri et al. [13] also used their dynamo model to calculate the variation of helicity with latitude over a solar cycle and found that the latitudinal distribution of helicity from their theoretical model is in broad agreement with observational data.

Accepting that the magnetic flux in the rising flux tube is nearly frozen, we expect that the poloidal flux collected by it during its rise through the SCZ should be confined in a narrow sheath at its outer periphery. In order to produce a twist in the flux tube,

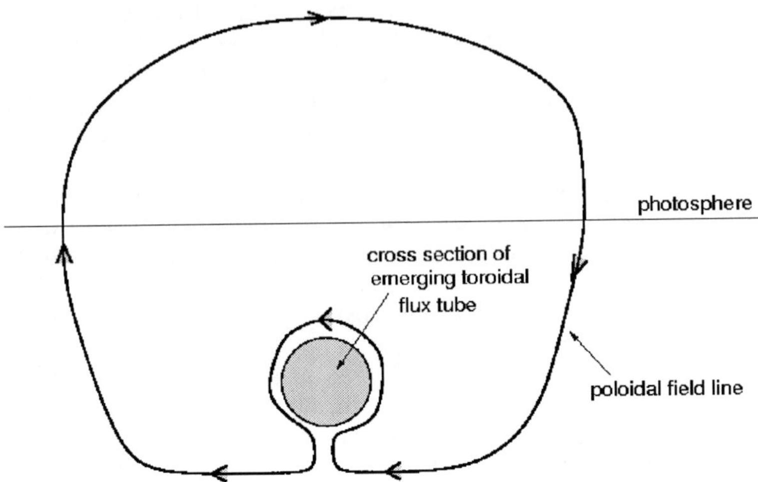

FIGURE 2. As a toroidal flux tube (here shown in cross section, hatched) rises through the convective zone, field lines of the weak external poloidal field may get wrapped around it.

the poloidal field needs to diffuse from the sheath into the tube by turbulent diffusion. However, turbulent diffusion is strongly suppressed by the magnetic field in the tube. This nonlinear diffusion process was studied in an untwisted flux tube by Petrovay and Moreno-Insertis [14]. The model was subsequently successfully applied for sunspot decay [15]. Chatterjee et al. [16] extended this model by including the poloidal component of the magnetic field (i.e. the field which gets wrapped around the flux tube) and studied the evolution of the magnetic field in the rising flux tube, as it keeps collecting more poloidal flux during its rise and as turbulent diffusion keeps acting on it.

The penetration of the wrapped up poloidal field into the flux tube can be modelled as follows. Consider a straight, cylindrical, horizontal magnetic flux tube rising through the solar convective zone. As all variables in this model depend only on the radial distance r from the tube axis and on time, we study the wrapping of the large-scale poloidal field around the flux tube by considering a radially symmetric accretion of azimuthal field by the flux tube. A further complication is the expansion of the flux tube during its rise, due to the decrease of the external pressure. This expansion is assumed to be self-similar, so that the Lagrangian radial coordinate ξ is related to the Eulerian radius r by $\xi = F(t)r$, where the expansion factor $F(t)$ was taken from thin flux tube emergence models. In the Lagrangian frame, flux density is rescaled as $B_z' = B_z/F^2$ and $B_\phi' = B_\phi/F$. With these notations and assumptions, the induction equation takes the form

$$\frac{\partial B_z'}{\partial t} = F^2 \frac{1}{\xi}\frac{\partial}{\partial \xi}\left(\eta \xi \frac{\partial B_z'}{\partial \xi}\right) \tag{3}$$

$$\frac{\partial B_\phi'}{\partial t} = F^2 \frac{\partial}{\partial \xi}\left[\eta \frac{1}{\xi}\frac{\partial}{\partial \xi}(\xi B_\phi')\right] - F\frac{\partial}{\partial \xi}(v B_\phi'). \tag{4}$$

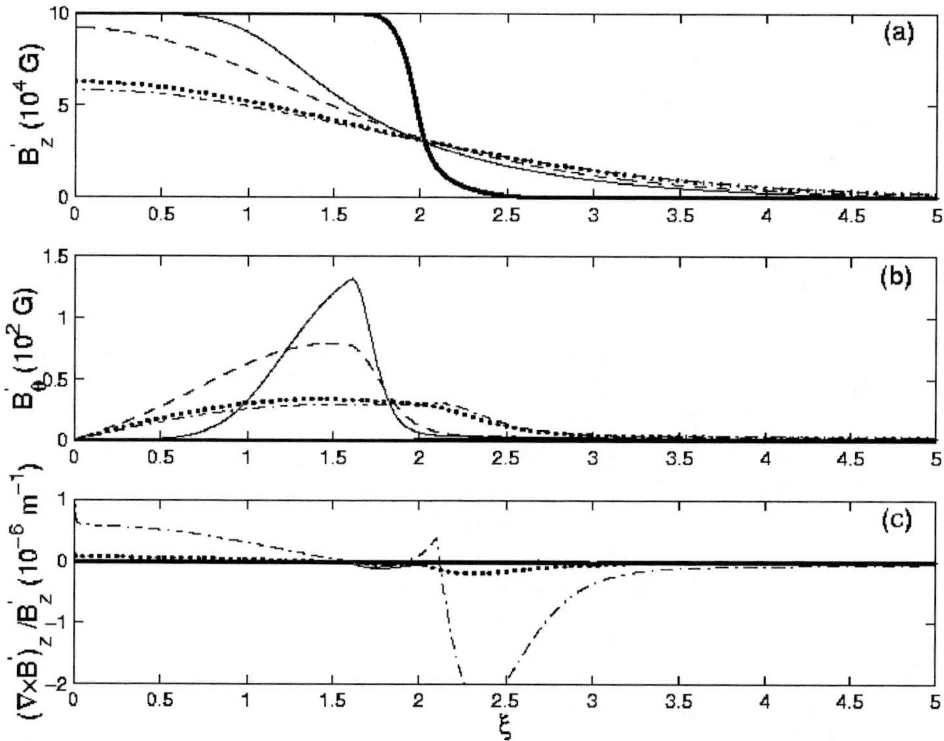

FIGURE 3. Plots of B'_z, B'_ϕ and α_p as functions of ξ for a rising flux tube (case A). The different curves correspond to the profiles of these quantities at the following positions of the flux tube: $0.7R_\odot$ (thick solid), $0.85R_\odot$ (solid), $0.9R_\odot$ (dashed), $0.95R_\odot$ (dotted), $0.98R_\odot$ (dash-dotted).

(Note that the advection term appears in the ϕ component only, as it does not represent a truly radially symmetric inflow; instead, it is just designed to mimick the effect of the wrapping of poloidal field lines around the rising tube.) Following Petrovay and Moreno-Insertis [14], the magnetic diffusivity η is specified in the form

$$\eta = \frac{\eta_0}{1+(B/B_{eq})^2} \tag{5}$$

Here η_0 is the magnetically unquenched diffusivity value at the scale of the flux tube. As the integral scale of turbulence L may be larger than this, η_0 is rescaled assuming a Kolmogorov spectrum:

$$\eta_0 = \eta_{00} \left(\frac{r_{ft}}{L} \right)^{4/3} \tag{6}$$

with

$$\eta_{00} = 3 \times 10^{12} \mathrm{cm}^2 \mathrm{s}^{-1}. \tag{7}$$

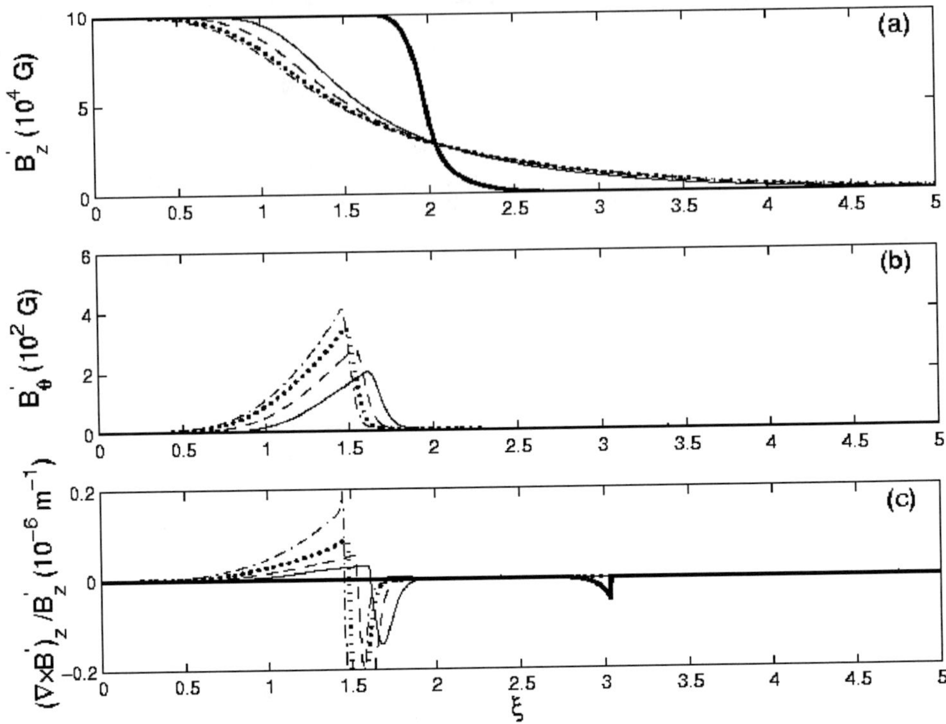

FIGURE 4. Same as Fig. 4 but the field inside the flux tube is not allowed to decrease below $3B_e$ at any height. (Case B)

All our calculations are performed for a flux tube of flux 10^{22} Mx and initial field strength 10^5 G.

The equations were numerically integrated by Chatterjee et al. [16]. The solutions are presented in Figures 3 and 4. The conclusions drawn from the model hinge on some assumptions, especially concerning the subsurface magnetic field structure in the last phases of the rise of the tube. The field strength in the rising tube, as calculated from thin flux tube emergence models, decreases well below the turbulent equipartition value B_e near the surface ($B_e^2/2\mu_0 = \rho\overline{v^2}/2$). The presence of 3000 G magnetic fields in sunspots is a compelling proof that magnetic fields may never fall to such low values; in fact, at least in photospheric layers, they are at about $3B_e$. This is presumably the result of flux concentration processes such as turbulent pumping and convective collapse. Thus, it may be more realistic not to allow the magnetic field to fall below $3B_e$. Figures 3 and 4 present results without and with this constraint, respectively.

From the lower panels of Figures 3 and 4 one finds that the typical value of α_p in the internal parts of the flux tube is of order $\sim 10^{-8}\,\mathrm{m}^{-1}$ at a depth of $0.85R_\odot$ in both cases. However, as the flux reaches the solar surface, in case A the B_ϕ component spreads out due to diffusion and its gradient becomes smaller, reducing α_p by about one order of magnitude. Only if the magnetic field inside the flux tube remains stronger than the equipartition field (the case B represented in Figure 4) is the B_ϕ component unable to diffuse inside so that its gradient remains strong and α_p is of order $\sim 10^{-8}\,\mathrm{m}^{-1}$ even near the surface. This suggests that our case B may be closer to reality, i.e. during the rise of the flux tube from $0.9R_\odot$ to $0.98R_\odot$ effective flux concentration processes are at work, keeping the field strength at a value somewhat above the equipartition level.

It should be added that in the calculations presented here the amplitude and sign of the poloidal field was assumed not to depend on depth. For alternative assumptions, significantly different current helicities may result, so the above conclusion should be treated with proper reservation. Details of the radial dependence of the poloidal field strength may strongly depend on the dynamo model. A more robust feature of the current helicity distributions, present in all the lower panels of our plots, is the presence of a ring around the tube with a current helicity of the opposite sense. This is clearly the consequence of the fact that on the outer side of the accreted sheath the radial gradient of the azimuthal field, and thus the axial current, is negative. This is an inevitable corollary of the present mechanism of producing twist in active regions. A rather strong prediction of this model is, therefore, that a ring of reverse current helicity should be observed on the periphery of active regions, somewhere near the edge of the plage.

4. WRAPPING OR RECONNECTION?

One precondition of the above scenario is that poloidal flux wrapping must take place around the rising tube before the inward diffusion of the field, modelled above, can give rise to a helical field structure. The implicit assumption behind the original proposal of Choudhuri [12] was that, as the magnetic Reynolds number is quite high in the solar plasma, freezing in can be considered valid to a good degree of approximation, making such wrapping inevitable. On the other hand, for the process of diffusion of the wrapped up field into the tube to take place as modelled above, freezing in must be violated. How can these two seemingly conflicting requirements be fulfilled simultaneously?

The key seems to be that they are *not* fulfilled simultaneously, but rather at different stages of the rise of the flux tube. The timescale of magnetic diffusion into the tube is the resistive timescale across the current sheet of thickness Δ surrounding the tube of radius R:

$$\tau_r = \Delta^2/\eta. \tag{8}$$

On the other hand, the timescale of poloidal field accretion is the rise timescale

$$\tau_c = R/V, \tag{9}$$

V being the velocity of the tube. So the ratio of the two timescales is

$$\tau_r/\tau_c = (\Delta/R)^2 \mathrm{Re_m}, \tag{10}$$

where $\mathrm{Re_m} = RV/\eta$ is the magnetic Reynolds number.

Figures 3 and 4 indicate that $\Delta \sim R$ during most of the rise (even though these computations were started from an initial state with a fairly thin current sheet, $\Delta(t=0) \sim$

$0.1R$). Plugging equations (5) and (6) into (10) we find that the ratio of the timescales is close to 10^2 in the early phases of the rise, near the bottom of the convective zone. By the time the tube reaches the uppermost 10–20 % of the convective zone, $\mathrm{Re_m}$ is reduced to less than 10, freezing in breaks down and resistive processes cannot be neglected any more. Thus, in the first 70 % or so of its path through the convective zone the tube sweeps up poloidal flux in more or less freezing in conditions, resulting in a wrapped up field with little infiltration into the tube. In the last phases of the rise the expanding and weakening tube cannot suppress turbulent motions any more, and the wrapped up field can diffusively penetrate the tube.

The assumed linear scaling (10) of the time scale ratio with the magnetic Reynolds number seems to be contradicted by the work of Linton and Antiochos [17]. In that work the collision of two twisted magnetic flux tubes was studied at different angles, and it was found that magnetic reconnection occurs at a rate determined by the collisional timescale (9). As a result, instead of wrapping around or bouncing off each other, the tubes reconnect. For low twist and/or collision angles not very different from perpendicular a simple reconnection occurs ("slingshot"), while for high twist and/or sharper collision angles repeated reconnection leads to the two tubes effectively "crossing" each other ("tunneling"). What is surprising is that these results seem to show very little sensitivity to the magnetic Reynolds number, even though values of $\mathrm{Re_m}$ up to 7200 were considered. Taken at face value, this may cast doubt on whether an external poloidal field can wrap around a rising toroidal tube for any value of the magnetic Reynolds number (Moreno-Insertis, private communication).

While in our scenario the flux tube interacts with a general background field instead of another flux tube, by analogy with Linton and Antiochos [17] we might still expect that instead of wrapping around the rising tube, the poloidal field will instantly reconnect to it. As here we are dealing with relatively low-twist fields and nearly perpendicular collision angles, the slingshot case should apply. If so, instead of wrapping this would simply result in some loss of toroidal flux from the tube, as it reconnects with the incoming poloidal field lines.

However, the root of the surprising behavior of the Linton and Antiochos [17] experiments is probably the rather artificial initial state used, where the flux tubes are bounded by an infinitely thin current sheet, i.e. $\Delta = 0$. At a time t after the start, the thickness of the sheet will have reached $\Delta \sim (\eta t)^{1/2}$. Substituting this into equation (8) above we find $\tau_r = t$: indeed, the reconnection will then keep step with the collision. It is straightforward to see that this conclusion is not changed if the resistivity has a turbulent origin and therefore itself scales with some power of Δ, as is the case in our model.

Dropping the assumption of a flux tube initially bounded by a discontinuity, the situation should change completely. As we have seen, in our models $\Delta \sim R$ during most of the rise even though we start from a state with a thinner current sheet. This suggests that the kind of wrapping necessary for the accretion of helicity may indeed take place after all, in the manner outlined above.

5. CONCLUSION

In the present paper, after briefly reviewing the available observational evidence on magnetic helicity in active regions, one particular mechanism was presented that is potentially capable of giving rise to such a twisted field structure. In addition to the accretion of poloidal fields by the emerging poloidal flux tube, described here, there are quite a few other subsurface mechanisms which may naturally introduce twist in the structure of emerging flux loops. Such proposed mechanisms include helicity generation by the solar dynamo [18] and buffeting of the rising flux tubes by helical turbulent motions [19]. A further possibility is the effect of Coriolis force on flows in rising flux loops [20]. As we mentioned above, this process is responsible for the generation of positive writhe in active region flux tubes in the northern hemisphere, and thereby for the observed tilt of active regions. As magnetic helicity is conserved in ideal MHD [21], the same process should then also give rise to a twist of opposite sense, so that the net helicity remains constant. The amount of helicity generated by this process, however, is too low to explain the observed values of α_p. On the other hand, a strongly twisted flux tube can develop a writhe by means of the kink instability. This mechanism should lead to a positive correlation of twist and writhe, in contrast to the helicity conservation argument above. Indeed, recent observations by López Fuentes et al. [22] and Holder et al. [23] indicate that this mechanism may be important in the case of at least some active regions with unusually high tilt.

An important clue to the origin of helicity in active regions could be yielded by a detailed mapping of the current helicity distribution in active regions. Future increase in magnetograph sensitivity may make such maps possible, thereby allowing direct testing of the predictions of some of theories mentioned above.

ACKNOWLEDGMENTS

This work was supported by the Theoretical Institute for Advanced Research in Astrophysics (TIARA) operated under Academia Sinica and the National Science Council Excellence Projects program in Taiwan administered through grant number NSC95-2752-M-007-006-PAE, by the Hungarian Science Research Fund (OTKA) under grant no. K67746 and by the European Commission through the SOLAIRE Network (MTRN-CT-2006-035484).

REFERENCES

1. G. E. Hale, *Nature* **119**, 708 (1927).
2. R. S. Richardson, *ApJ* **93**, 24–28 (1941).
3. J. B. Zirker, S. F. Martin, K. Harvey, and V. Gaizauskas, *Solar Phys.* **175**, 27–44 (1997).
4. N. Seehafer, *Solar Phys.* **125**, 219–232 (1990).
5. A. A. Pevtsov, R. C. Canfield, and T. R. Metcalf, *ApJ Lett.* **440**, L109–L112 (1995).
6. K. D. Leka, and A. Skumanich, *Solar Phys.* **188**, 3–19 (1999).
7. A. B. Burnette, R. C. Canfield, and A. A. Pevtsov, *ApJ* **606**, 565–570 (2004).
8. L. van Driel-Gesztelyi, P. Démoulin, and C. H. Mandrini, *Advances in Space Research* **32**, 1855–1866 (2003).

9. S. D'Silva, and A. R. Choudhuri, *A&A* **272**, 621–633 (1993).
10. C. R. DeVore, *ApJ* **539**, 944–953 (2000).
11. A. A. Pevtsov, V. M. Maleev, and D. W. Longcope, *ApJ* **593**, 1217–1225 (2003).
12. A. R. Choudhuri, *Solar Phys.* **215**, 31–55 (2003).
13. A. R. Choudhuri, P. Chatterjee, and D. Nandy, *ApJ Lett.* **615**, L57–L60 (2004).
14. K. Petrovay, and F. Moreno-Insertis, *ApJ* **485**, 398–408 (1997).
15. K. Petrovay, and L. van Driel-Gesztelyi, *Solar Phys.* **176**, 249–266 (1997).
16. P. Chatterjee, A. R. Choudhuri, and K. Petrovay, *A&A* **449**, 781–789 (2006).
17. M. G. Linton, and S. K. Antiochos, *ApJ* **625**, 506–521 (2005).
18. N. Seehafer, M. Gellert, K. M. Kuzanyan, and V. V. Pipin, *Advances in Space Research* **32**, 1819–1833 (2003).
19. D. W. Longcope, G. H. Fisher, and A. A. Pevtsov, *ApJ* **507**, 417–432 (1998).
20. Y. Fan, and D. Gong, *Solar Phys.* **192**, 141–157 (2000).
21. M. Berger, *Plasma Phys. Contr. Fusion* **41**, B167–B175 (1999).
22. M. C. López Fuentes, P. Démoulin, C. H. Mandrini, A. A. Pevtsov, and L. van Driel-Gesztelyi, *A&A* **397**, 305–318 (2003).
23. Z. A. Holder, R. C. Canfield, R. A. McMullen, D. Nandy, R. F. Howard, and A. A. Pevtsov, *ApJ* **661**, 1149–1159 (2004).

The Origin of Solar Eruptions

T. G. Forbes

Institute for the Study of Earth, Oceans, and Space, University of New Hampshire, Durham, NH 03824, USA

Abstract. The occurrence of eruptions in the solar atmosphere has been known for nearly a 150 years, yet the underlying mechanism that creates them remains obscure. Most present-day models are based on the principle that the energy that drives an eruption comes from the magnetic energy associated with stressed magnetic fields in the solar corona. However, there is no general agreement in the solar research community as to what triggers the release of this energy. One possibility is that it is caused by a combination of ideal (e.g. loss of equilibrium) and non-ideal (e.g. magnetic reconnection) processes. The first process can explain the rapid onset of the eruption, but the second is needed to explain the large scale of the energy release.

Keywords: coronal mass ejection, solar flare, magnetic reconnection
PACS: 96.60.qe, 96.60.ph, 96.60.qf, 96.60.Iv

INTRODUCTION

In this paper we use the term "eruption" to refer to phenomena such as a flare, a coronal mass ejection (CME) or a prominence eruption. These phenomena are all closely related and are probably different manifestations of a single, underlying physical process, namely, the release of magnetic energy stored in the magnetic field of the solar atmosphere.

Historically, a flare was defined as a localized brightening of the chromosphere, observed in H-alpha, over a time scale ranging from a few minutes to an hour [1]. Nowadays, the definition has been extended to include the rapid onset of X-ray and UV emissions in the corona [2]. The H-alpha brightening in the chromosphere typically occurs in the form of flare ribbons, while the X-ray and UV emissions appear in the form of loops whose feet map to the ribbons. During the course of the flare the separation between the ribbons typically increases, and the flare loops grow in size.

CMEs correspond to an expulsion of mass from the low corona in interplanetary space. They typically have a three-part structure consisting of an outer bright shell, an interior dark cavity, and prominence material located near the center of the cavity [3]. The density of the shell is about a factor of 10 higher than the ambient coronal density. This value is too large for the enhanced density region to be produced by shock compression. The shell most likely results from the pile-up of the material of the helmet streamer that typically overlies the erupting region [4]. The outward propagating CME does produce a shock, but this shock is difficult to observe. In those cases where the shock has been detected, the density compression across it is typically in the range from 1.2 to 2.5 according to Vourlidas [5].

More than half of all CMEs are associated with the eruption of prominences. Large quiescent prominences outside active regions tend to create slow to moderate speed

CP934, *Flows, Boundaries, Interactions*
edited by C. Dumitrache, V. Mioc, and N. A. Popescu

CMEs. High-speed CMEs are typically produced in active regions, and they are generally well correlated with flares. As observations have improved, it has become increasingly clear that erupting prominences outside active regions have many features typical of large flares. Like large flares, erupting prominences produce loops and ribbons that move apart in time, but, unlike large flares, the ribbons are usually too faint to be seen in H-alpha. However, the ribbons can often be seen in the He 10830 Å line that is a more sensitive indicator of chromospheric excitation [6]. The eruption of a large quiescent prominence does not usually produce significant hard X-ray emissions, probably because it occurs in a region where the field is relatively weak (< 10 Gauss).

From time to time there has been considerably controversy about the relation between CMEs and flares. Some authors have argued that flares cause CMEs, but most CMEs are not associated with what is normally considered a flare [7]. Flares occur over a span of energy scales that range from very small (microflares and nanoflares) to very large (> 10^{32} ergs), and the small ones in particular are far more numerous than CMEs. It is the large flares that tend to be associated with CMEs. Sometime ago, Svestka and Cliver [8] suggested that the main factor that determines whether a CME will be associated with a flare or not, is the strength of the magnetic field in the erupting region. If the ambient magnetic field strength is weak, then the emitted radiation, although still present, is just too faint to be considered a flare. This has been explicitly demonstrated by Reeves and Forbes [9] who have shown that it is theoretically possible to have two CMEs with nearly the same trajectories and speeds but with an order of magnitude or more difference in the peak intensities of their light curves. A low mass CME in a weak field region can experience the same acceleration as high mass CME in a strong field region, but the reduced magnetic energy density of the former case leads to much weaker emissions.

CME associated flares usually constitute what are known as long duration events (LDEs), and they can last for longer than 10 hours. By contrast, small non-CME associated flares may last 30 minutes, or less, depending on their size. The difference in the time scale can be understood as a consequence of the very large length scale created by the CME. A compact magnetic field configuration that becomes unstable or loses equilibrium can quickly reach a new equilibrium state if it remains confined. However, if the field erupts into a large-scale structure extending many solar radii outward into the weak field region of interplanetary space, it takes a long time for the field to reach a new equilibrium state.

A TWO-DIMENSIONAL ANALYTICAL MODEL

About 30 years ago Van Tend and Kuperus [10] suggested that eruptive phenomena such as coronal mass ejections, prominence eruptions, and large flares might be caused by a loss of equilibrium in a coronal flux rope. Since then, there has been an ongoing effort to develop increasingly sophisticated models that incorporate their basic idea (e.g. Molodenskii and Filippov [11], van Ballegooijen and Martens [12], Forbes and Isenberg [13]). When equilibrium is lost, the flux rope rises rapidly, and, as it rises, a current sheet forms below it. In the two-dimensional models, rapid reconnection is required in the current sheet in order for the flux-rope to escape from the Sun into interplanetary space.

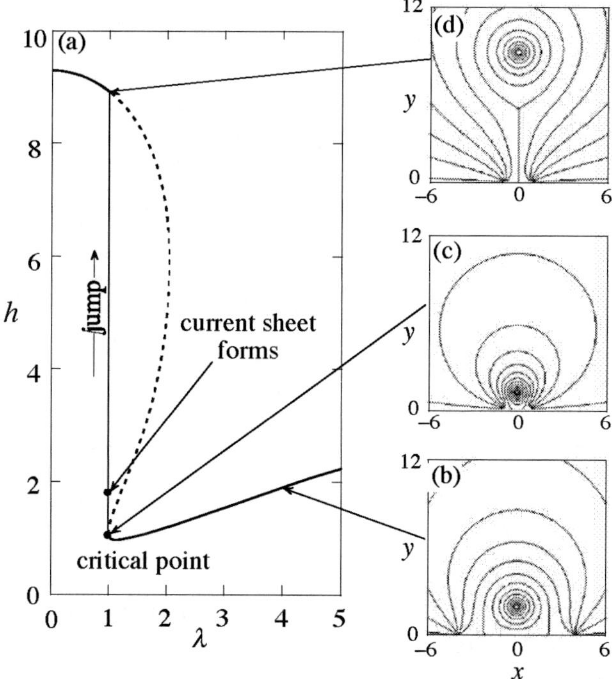

FIGURE 1. A two-dimensional, analytical model for the onset of solar eruptions. The left panel (a) shows the equilibrium height, h, of the flux rope as a function of the separation half-distance, λ, between the photospheric sources at $y = 0$. The dashed section of the curve indicates the region where the equilibria are unstable. If λ is slowly decreased, the configuration will reach a critical point where the flux rope suddenly jumps upwards. Figures (b), (c), and (d) show contours of the flux function $A(x,y)$ at the three locations indicated in (a). All distances are in normalized units, and the initial flux rope radius is 0.1 (after Forbes and Priest[16].

In the absence of significant pressure gradients and gravity, two-dimensional equilibrium configurations in the corona are described by the Grad-Shafranov equation [14, 15]

$$\nabla^2 A + \frac{1}{2}\frac{dB_z{}^2}{dA} = 0$$

in the semi-infinite $x - y$ plane with $y > 0$, where B_z is the field perpendicular to this plane and $A(x,y)$ is the flux function. The surface at $y = 0$ corresponds roughly to the photosphere. The Grad-Shafranov equation can be used to construct an evolutionary sequence of force-free equilibria in response to quasi-statically slow changes in the photospheric boundary conditions. That is, changes which occur on time scales much greater than the time for magneto-acoustic waves to cross the configuration.

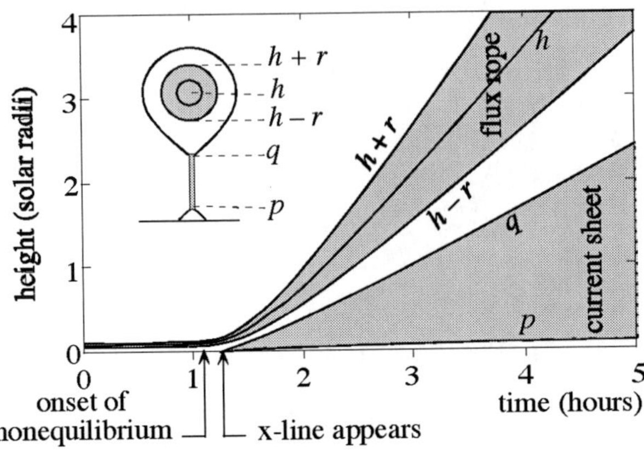

FIGURE 2. Flux rope and current sheet trajectories obtained from the model in Figure 1 by assuming a constant inflow Alfvén Mach number of 0.1 in the inflow region at the midpoint of the current sheet. The parameters h, r, q, and p are the flux rope's height and radius and the current sheet's upper and lower tips, respectively.

Figure 1 shows a model based on these principles. The magnetic field of the model is prescribed in terms of the flux function A as

$$A(x,y) = \mathrm{Re}\left[\frac{2I}{c}\ln\left(\frac{\sqrt{\zeta^2+q^2}+i\sqrt{h^2-q^2}}{\sqrt{\zeta^2+q^2}-i\sqrt{h^2-q^2}}\right)+i\frac{A_0}{\pi}\ln\left(\frac{\sqrt{\zeta^2+q^2}+\sqrt{q^2+\lambda^2}}{\sqrt{\zeta^2+q^2}-\sqrt{q^2+\lambda^2}}\right)\right]$$

for $(\zeta - ih) > r$, where $\zeta = x + iy$ is the complex coordinate, r is the flux-rope radius, c is the speed of light, I is the flux rope current, q is the height of the current sheet below the flux rope, and A_o is the value of A at the origin. The panels on the right of figure 1 show contours of A for different values of λ.

The current, I, is determined from the application of the frozen-flux condition at the surface of the flux rope. This is done by setting $A(0,h-r) =$ constant. The radius r is determined by applying the frozen flux condition to the interior field of the flux rope [16]. The height, q, of the tip of the current sheet is determined by requiring the magnetic field their to have a y-type null-point.

Regardless of the value of the flux rope height, h, this model always satisfies the same boundary condition, namely

$$A(x,0) = A_0 \mathscr{H}(\lambda - |x|),$$

where $\mathscr{H}$ is the Heavyside step function. This boundary condition corresponds to two point sources of opposite polarity located at $x = \pm\lambda$, and its invariance with regards to h means that the magnetic field is line-tied during an eruption. That is, no magnetic flux or energy can enter through the lower surface during an eruption.

In the absence of magnetic reconnection, the flux rope in the model cannot escape. If one places the flux rope at the critical point, it will start to move upward, slowly at first, but with an ever increasing speed until it reaches a height corresponding to $h = 9\lambda$. At this height it start to slow down, and it will eventually come to rest at this location if its kinetic energy is dissipated (cf. Figure 1d). However, if reconnection is allowed, the flux rope will escape provided that the reconnection is fast enough. A version of the model that allows for such reconnection has been developed by Lin and Forbes [17]. They found that even a very small amount of reconnection is sufficient to allow the flux rope to escape smoothly without any deceleration. For reconnection rates corresponding to an inflow Alfvén Mach number, M_A, > 0.05 (at the midpoint of the current sheet sides) the flux rope could still escape without any deceleration. Escape with deceleration occurred as long as M_A was larger than 0.006. Figure 2 shows an example of a non-decelerating trajectory obtained for $M_A = 0.1$. As the flux rope moves upward, the current passing through it decreases as required by the conservation of magnetic flux. The lower tip of the current sheet at p moves upward very slowly because its motion is controlled by the slow rate at which reconnection occurs. By contrast, the upper tip of the current sheet at q moves upward quite rapidly at a speed that is only about a factor of two smaller than the speed of the flux rope. Because a loss-of-equilibrium is an ideal-MHD process, the upward speed of the flux rope is close to the ambient Alfvén speed in the region near the base.

MAGNETIC ENERGY RELEASE

In this particular model an x-line does not appear until the flux rope reaches a height of about $\sqrt{3}$ times its initial height at the critical point. Once the x-line appears, a current sheet forms, and it is possible to calculate the Poynting flux of magnetic energy through the sheet. A representative result is shown in Figure 3. The Poynting flux supplies the magnetic energy that powers the flare ribbons and loops. For the case shown, less than 20% of the total magnetic energy released is channeled through the current sheet. However, for smaller values of M_A, the percentage increases until at $M_A = 0.01$ it is nearly 80%.

The analytical model used to make Figure 3 approximates the reconnection region as a simple current sheet, and it assumes that the reconnection rate is given. The model also assumes that all of the magnetic energy flowing into the reconnection region is eventually thermalized and that half of this amount is transferred into the flare loop system. The other half is transferred upward. Two-dimensional numerical simulations [18, 19] suggest that actual situation is not quite so simple. First, most of the thermal plasma in the flare loops is probably created when energetic particles generated directly, or indirectly, by reconnection impact the lower chromosphere. Second, some fraction of the magnetic energy released in the reconnection region will not be thermalized, but will, instead, end up contributing to the acceleration of the flux rope. Third, it is likely that more energy is channeled upwards than downwards with the amount varying in time. Thus, the analytical model probably overestimates the fraction of the magnetic energy that is channeled into the flare loop system.

Both the total magnetic power output and the Poynting flux rise rapidly on a time scale that is of the same order as the Alfvén time scale of the system. After reaching a peak value of about 5×10^{28} ergs/s, the current sheet power declines. The decline occurs

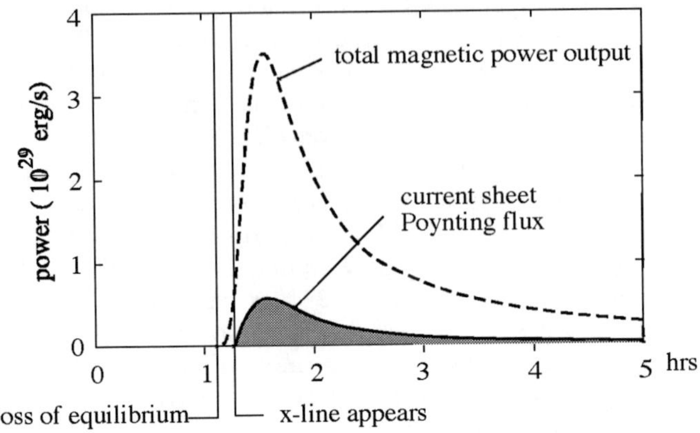

FIGURE 3. The predicted magnetic energy release as a function of time for the same case shown in Figure 2. In this case, less than 20% of the total magnetic energy released is channeled through the reconnection region as indicated by the curve labeled "current sheet Poynting flux".

quickly at first, due to the initially rapid upward motion of the x-line. However, after about two hours, the decline becomes very gradual which reflects the very slow time scale of the reconnection process at high altitudes.

THE PREDICTED EVOLUTION OF SPECTRAL LINES

The Poynting flux passing into the current sheet can be used as an input for a multi-loop numerical model of the flare loop system. At any given time there are only a small number of field lines that map to the current sheet (and/or slow shocks). Conduction electrons and energetic particles travel along these field lines down to the chromosphere and result in an impulsive heating of the plasma as shown in Figure 4. Some of the heated plasma then flows back up into the corona to produce the system of flare loops shown in the figure [20]. Once a magnetic field line is disconnected from the current sheet, the plasma on it cools, first by thermal conduction and later by radiation [21]. This process leads to the formation of a system of flare loops with different temperatures, densities, and flows.

Because the amount of time that a given field line is connected to the reconnection region is short, it is possible to model the overall emission of the loop system as a collection of many small loops heated at different times by a continually evolving source. Models of this type have been constructed by Hori et al. [22, 23], Reeves and Warren [24], Warren and Doschek [25], and Warren [26]. These models have used either an empirical energy input derived from data or an arbitrary heating function specifically tailored to match the observations. The models show that a sustained energy input is required to account for the emissions produced in LDEs. Recently, Reeves et al. [27] have used the reconnection energy output predicted by the two-dimensional model shown in Figures 2 and 3 to drive a multi-loop model. With the use of the

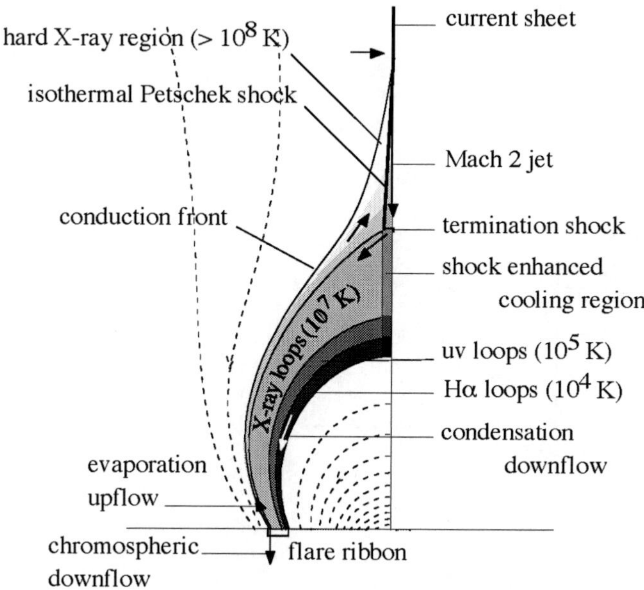

FIGURE 4. Diagram of the temperature structure and flows predicted by a reconnection model of flare loops. Strong thermal conduction channels the energy released by reconnection to the chromosphere where it heats the plasma. The high pressure thereby created drives plasma upwards into the corona and downwards into the lower corona (from Forbes and Acton [20]).

SOLFTM one-dimensional flare code [28] they obtained predictions for the density, temperature, and bulk flow of the entire flare loop system as functions of space and time. This information was then processed through a numerical routine (bcs-spec, part of the standard SolarSoft package) developed at the US Naval Research Laboratory for the Bragg Crystal Spectrometer (BCS) on the Japanese satellite Yohkoh. Figure 5 shows the resulting spectrum for the CaXIX line at three different times.

The model's parameters were chosen to simulate a B2.5, LDE flare. Even though the flare generated by this model consists of many loops containing flowing plasma at different temperatures, the combined spectrum that is generated is similar to the spectrum generated by a single loop at a temperature of 8 MK at the time of the peak emission in CaXIX. The model predicts that during the early phase of the flare a strong blue shift occurs, but only when the intensity of the line is still below the intensity threshold of the instrument. By the time emission becomes detectable, the spectrum consists of a stationary component with an enhanced blue wing. By the time of the peak emission, only the stationary component is evident. These results nicely account for the observations of actual flares by the BCS [29, 30].

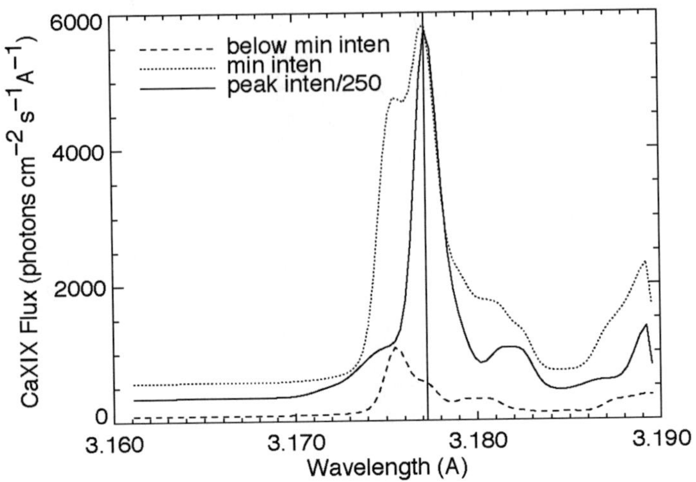

FIGURE 5. The CaXIX line profile predicted by the model shown in Figures 2 and 3. Three different times are shown corresponding to the peak intensity (re-scaled solid line), an earlier time when the intensity would first be observed by the Yohkoh BCS (dashed line), and an even earlier time when the intensity would be too low to be observed (dotted line). The vertical line marks the rest wavelength of the resonance line (from Reeves et al. [27]).

FUTURE DIRECTIONS

The two-dimensional loss-of-equilibrium model outlined in the previous sections accounts for many of the basic properties of flares and CMEs. However, many questions remain to be answered. For example, how are the currents in the corona created, and what form do they take in a three-dimensional configuration? Can models of the above type account for the acceleration of energetic protons and electrons? What kind of interaction occurs between the erupted field and the surrounding non-erupted field and can such interactions be detected? How would such a model work in three-dimensions? Would a more elaborate model be able to account quantitatively for actual observations? The questions that one can ask of this sort are almost limitless. At the present time, many researchers are actively involved in trying to supply answers.

ACKNOWLEDGMENTS

The author wishes to thank Dr. Cristiana Dumitrache, Dr. Nedelia A. Popescu, and Dr. Vasile Mioc for their hospitality. He also would like to thank them for the excellent IHY workshop and summer school that they organized in Sinaia. The research reported in this paper was supported by NSF grants ATM-0422764, ATM-0518218, ATM-0519249 and NASA grants NNX-06AC19G, NNH-05AA131 to the University of New Hampshire. The author's travel to this workshop was supported by the Romanian Academy of Science.

REFERENCES

1. H. Zirin, *Astrophysics of the Sun*, Cambridge Univ Press, New York, 1988.
2. E. Tandberg-Hanssen, and A. G. Emslie, *The Physics of Solar Flares*, Cambridge Univ. Press, New York, 1988.
3. B. C. Low, *Solar Phys.* **167**, 217–265 (1996).
4. A. Hundhausen, "The origin and propagation of coronal mass ejections," in *Proceedings of the Sixth International Solar Wind Conference*, edited by V. Pizzo, T. Holzer, and D. Simes, NCAR/TN 306, Boulder, 1988, pp. 181–241.
5. A. Vourlidas, "Detections of CME-driven shocks with LASCO," in *SOHO-17: 10 Years of SOHO and Beyond*, edited by H. Lacoste, and L. Ouwehand, European Space Agency, Nordwiijk, 2006, vol. ESA SP-617, p. 23.1.
6. K. Harvey, and F. Recely, *Solar Phys.* **91**, 127–139 (1984).
7. J. Gosling, *J. Geophys. Res.* **98**, 18937–18949 (1993).
8. Z. Svestka, and E. Cliver, "History and basic characteristics of eruptive flares," in *Eruptive Solar Flares*, edited by Z. Svestka, B. Jackson, and M. Machado, Springer-Verlag, New York, 1992, pp. 1–14.
9. K. Reeves, and T. Forbes, *Astrophys. J.* **610**, 1133–1147 (2005).
10. W. Van Tend, and M. Kuperus, *Solar Phys.* **59**, 115–127 (1978).
11. M. Molodenskii, and B. Filippov, *Soviet Astron. (English. Translation)* **31**, 564–568 (1987).
12. A. van Ballegooijen, and P. Martens, *Astrophys. J.* **343**, 971–984 (1989).
13. T. G. Forbes, and P. Isenberg, *Astrophys. J.* **373**, 294–307 (1991).
14. H. Grad, and H. Rubin, "MHD equilibrium in an axisymmetric toroid," in *Proceedings of the 2nd UN Conf. on the Peaceful Uses of Atomic Energy*, IAEA, Vienna, 1958, vol. 31, p. 190.
15. V. Shafranov, *Revs. of Plasma Phys.* **2**, 103–151 (1966).
16. T. Forbes, and E. Priest, *Astrophys. J.* **446**, 377–389 (1995).
17. J. Lin, and T. Forbes, *J. Geophys. Res.* **105**, 2375–2392 (2000).
18. T. G. Forbes, J. M. Malherbe, and E. R. Priest, *Solar Phys.* **120**, 285–307 (1989).
19. T. Yokoyama, and K. Shibata, *Astrophys. J. Letts.* **549**, 1160–1174 (2001).
20. T. Forbes, and L. Acton, *Astrophys. J.* **459**, 330–341 (1996).
21. P. Cargill, J. Mariska, and S. Antiochos, *Astrophys. J.* **439**, 1034–1043 (1995).
22. K. Hori, T. Yokoyama, T. Kosugi, and K. Shibata, *Astrophys. J.* **489**, 426–441 (1997).
23. K. Hori, T. Yokoyama, T. Kosugi, and K. Shibata, *Astrophys. J.* **500**, 492–506 (1998).
24. K. Reeves, and H. P. Warren, *Astrophys. J.* **578**, 590–597 (2002).
25. H. P. Warren, and G. A. Doschek, *Astrophys. J.* **618**, L157–L160 (2005).
26. H. P. Warren, *Astrophys. J.* **637**, 522–530 (2006).
27. K. Reeves, H. Warren, and T. Forbes, *Astrophys. J.* p. in press (2007).
28. J. T. Mariska, *Astrophys. J.* **319**, 465–480 (1987).
29. E. Antonucci, M. A. Dodero, G. Peres, S. Serio, and R. Rosner, *Astrophys. J.* **322**, 522–543 (1987).
30. G. A. Doschek, and H. P. Warren, *Astrophys. J.* **629**, 1150–1163 (2005).

Eruptive and Compact Flares

B.Schmieder*, G.Aulanier*, C.Delannée† and A.Berlicki**

*LESIA, Observatoire de Paris, 92195, Meudon, France
†Royal Observatory of Belgium, Brussels, Belgium
**Astronomical Institute of the University of Wroclaw, Wroclaw, Poland

Abstract. Solar two ribbon flares are commonly explained by magnetic field reconnections in the high corona. During the reconnection energetic particles (electrons and protons) are accelerated from the reconnection site. These particles are following the magnetic field lines down to the chromosphere. As the plasma density is higher in these lower layers, there are collisions and emission of radiation. Thus after the flare bright ribbons are observed at both ends of loops. These ribbons are typically observed in Hα and in EUV with SoHO and TRACE. As the time is going, these ribbons are expanding away of each other. In most studied models, the reconnection site is an X-point, where two magnetic separatrices intersect. They define four distinct connectivity domains, across which the magnetic connectivity changes discontinuously. In this paper, we present a generalization of this model to 3D complex magnetic topologies where there are no null points, but quasi-separatrices layers instead. In that case, while the ribbons spread away during reconnection, we show that magnetic field lines can quickly slip along them. We propose that this new phenomenon could explain fast extension of Hα and TRACE 1600 Å ribbons, fast moving HXR footpoints as observed by RHESSI, and that it is observed in soft X rays with Hinode/XRT. We also show how this concept can be applied to model the non-wave nature of EIT waves.

Keywords: Solar flares, magnetic field, MHD simulation
PACS: 95.85.Kr,95.85.Nv,96.60.Iv,96.60.Mz,96.60.Na,96.60.P,96.60.qe,96.60.Xy

INTRODUCTION

Eruptions occur at locations on the Sun where there is a build-up of a large amount of magnetic energy and eruptions occur when this energy is released via some reconnection process. Many models deal with how this energy release is accomplished for example [1] and [2], discuss how convergence of magnetic flux below a flux tube situated along a magnetic polarity inversion line can lead to enhancement of the flux tube and eruption, [3] present a model where eruption results when the tension of the sheared arcade or flux tube field is weakened via reconnection between the main magnetic field and emerging flux (kind of "tether-cutting" mechanism). [4] discuss instead disruption of field lines and reconnection above flux tubes. To determine which of the mechanism is predominantly responsible for eruptions it is necessary to observe the onset and the evolution of flares. The coronal plasma is frozen into the magnetic field almost everywhere except where current sheets can be formed and then dissipated. Current sheets develop along separatrices when the magnetic configuration evolves quasi-statically or dynamically. A very powerful tool to understand where the energy could be deposited is to study the magnetic topology of the active region, since it defines where magnetic reconnection is expected to occur (see reviews of [5], [6], [7]).

In the first section, magnetic topology models are presented. In the second and third

CP934, *Flows, Boundaries, Interactions*
edited by C. Dumitrache, V. Mioc, and N. A. Popescu

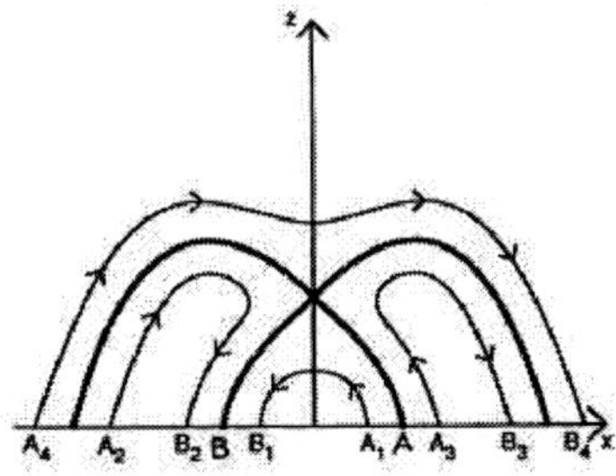

FIGURE 1. 2D magnetic reconnection sketch in a quadrupolar configuration ([8])

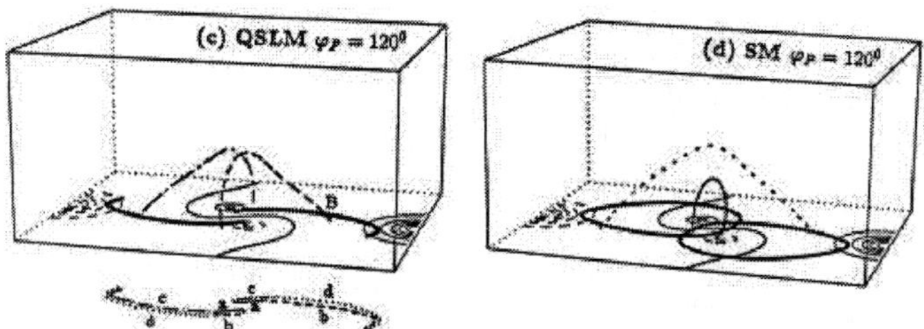

FIGURE 2. 3D magnetic models with four charges: (right panel) separatrice case; (left panel) quasi-separatrice layer (QSL) case. The ribbons are reduced in length in the QSL case and better comparable with the observations ([9]).

sections, We review eruptive and compact flares showing that their dynamics cannot be neglected to interpret the formation of long ribbons with evolving Hα, UV brightenings and X-ray kernels. New simulations are briefly described. Slip-running reconnection simulation explains the fast extension of ribbons; inflating separatrices simulation after flares allows us to interpret EIT waves as moving footprints of current shells in the atmosphere.

MAGNETIC TOPOLOGICAL MODELS

Coronal activities such as flares, eruptions and general heating are often attributed to the manner in which the coronal field responds to photospheric motion. The coronal plasma has a low resistivity and to release efficiently the free energy, small length scale have to be created. Coronal magnetic fields are forced to evolve continuously by slow photospheric velocities (0.1 km/s), faster velocities still lower to the typical Alfvén velocity ($\sim$ 1000 km/s) may exist during magnetic flux emergence. In this context magnetic configurations, with a slow evolution at the boundary leads to the formation of very thin current layers play a key role (review [10]).

Let us recall some classical definitions of magnetic configuration and reconnection. Figure 1 represents the models in 2D of separatrices existing in quadrupolar configuration. Separatrices are magnetic surfaces where the magnetic field line linkage is discon-

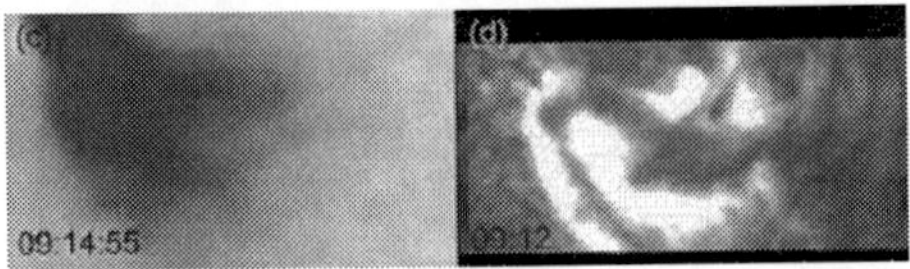

FIGURE 3. October 27, 1993 flare (a) in X ray (Yohkoh) and (b) in Hα (Pic du Midi)([11]).

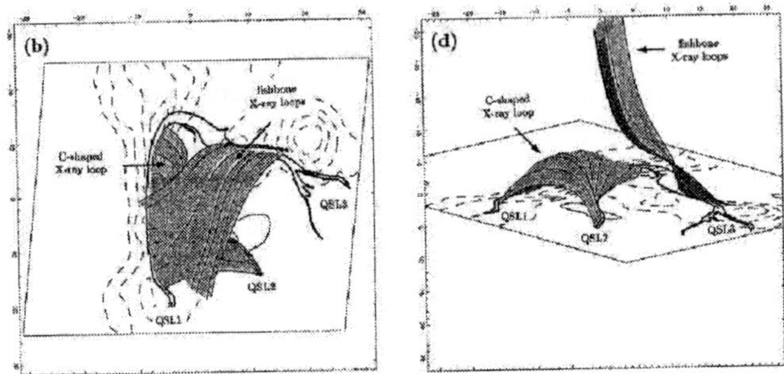

FIGURE 4. lfff extrapolation of the flare observed on the October 27, 1993 showing the QSL locations (contours) fitting with the Hα ribbons (b) before the flare, (d) after the flare ([11]).

tinuous, thus where the field lines can change abruptly of connectivity. In 2D configuration the reconnection occurs at X-point (null point) where the magnetic vanishes, in 3D configuration the intersection of the field lines occur along the separator, intersection of the separatrices. Current sheet form along the whole separatrices when shearing flows are present in the photosphere. The flare ribbons are found at the intersections of the separatrices and the chromosphere (Figure 2 right panel). Flare ribbons and loops evidence the locations where the stored energy is released. For testing the spatial localization of the energy release sites, it is needed to extrapolate the field lines in the corona using photospheric magnetograms as a boundary.

When the magnetic field is no more represented by point charges but maintained continuously as observed, there are not necessarily separatrices to interpret the ribbons: a generalization of the concept of separatrices does provide an interpretation of 3D magnetic reconnection. [12],[9] introduce this new concept based on the existence of quasi-separatrice layers (QSL) for general magnetic fields anchored to a boundary. In Figure 2 (left panel) we see that QSLs are defined as volumes where magnetic field lines can change of connectivity abruptly. There are no coronal null points in such configurations. such topological properties are insensitive to detailed geometry of the magnetic field, and thereby create a very robust tool to understand the loci of flare energy release. The ribbons are overlaid to the intersection of QSLs and the chromosphere, they are restricted to smaller regions than in the case of separatrices and match better the shape of the ribbons. We illustrate this point with the Figures 3 and 4, where observations and quasi-separatrices are shown ([11]).

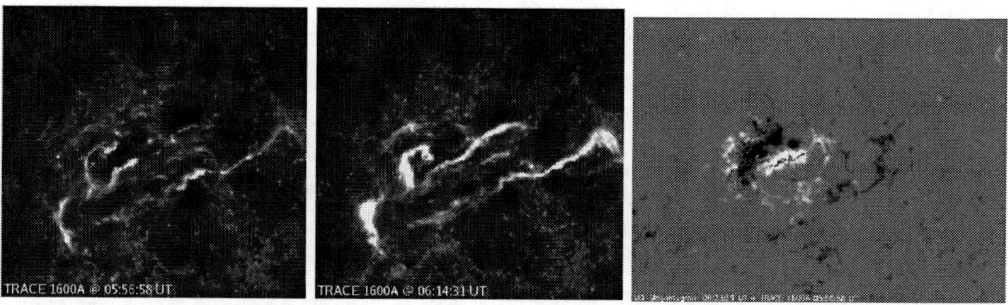

FIGURE 5. Flare ribbon fast extension observed on May 27, 2003, observed by TRACE in 1600 Å (left panels), complex magnetic configuration of the active region (MDI) overlaid by flare ribbons (courtesy of A. Bérurier)

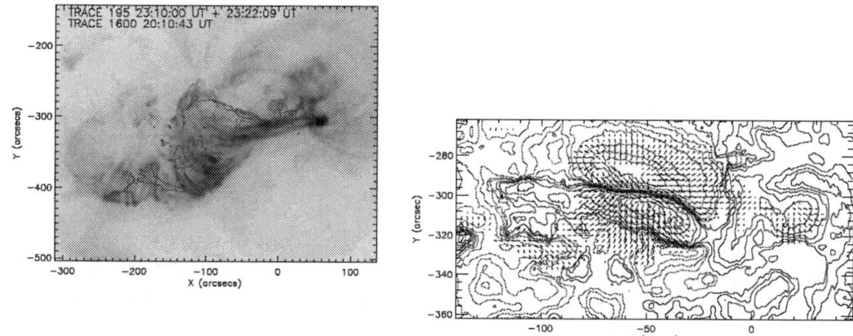

FIGURE 6. Complex active region registering a series of 10 flares in 12 hours on September 13 2005, 9left panel) example of flare ribbons observed during the three last flares at, 20:10 UT, 23:10 UT and 23:22 UT (courtesy of H. Li), (right panel) photospheric vector magnetic field observed with THEMIS on September 13, 2003 (courtesy of V. Bommier)

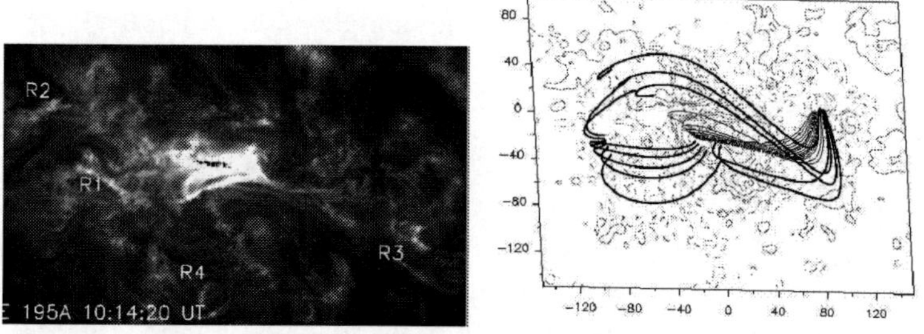

FIGURE 7. Pre-event of the X 17 flare occurring on October 28, 2003 showing four ribbon in TRACE 1600 Å (left panel), extrapolation of the magnetic field lines showing the change of connectivity with the four ribbons R1, R2, R3, R4 (right panel), the field lines are anchored in the four ribbons suggesting a magnetic reconnection on large scale, no null point was found. The bright loops in the middle of the active region concern a low reconnection which takes place during a long time as a new bipole was emerging. This small event does not interact with the large scale quadrupolar reconnection ([13])

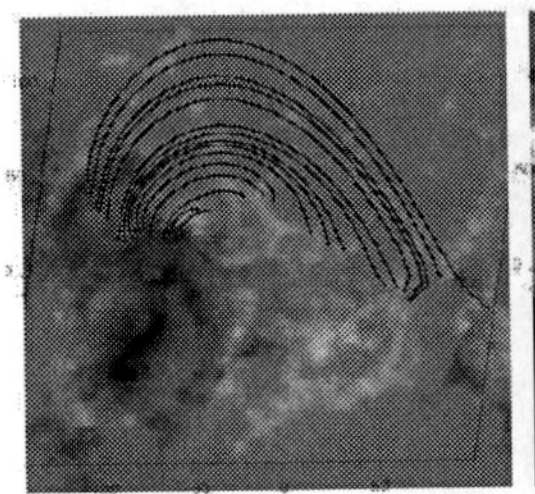

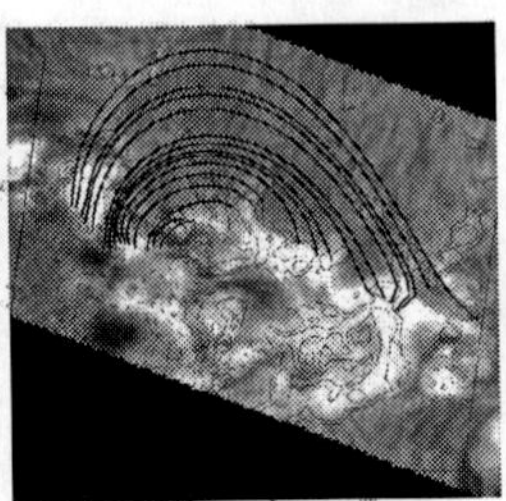

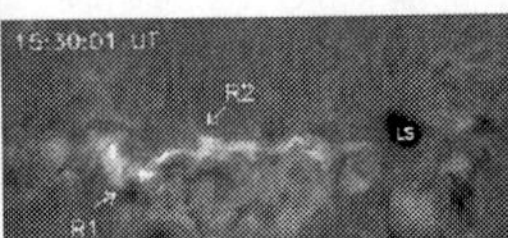

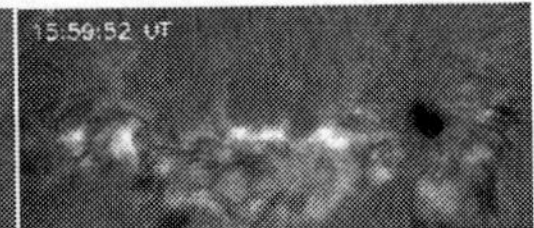

FIGURE 8. Magnetic field lines extrapolated in a linear force free configuration and Hα ribbons of the 22 October 2003 flare for two times observed with the VTT in Canaries showing the fast extension of the ribbons ([23])

ERUPTIVE FLARES

Firstly we investigated the regions of large flares, we called eruptive flares when a coronal mass ejection (CME) is related to the flare. In this frame, [3] and [14] proposed unified models. As the magnetic field lines are reconnected, a plasmoid is escaping and forming the CME. During the reconnection, electrons are precipiced downwards to denser layers. They heat the plasma and form the so-called two bright ribbons in the chromosphere (visible in Hα, in Ca II H by Hinode/SOT, the 1600 Å continuum by TRACE), in the transition region (CDS, EIT and TRACE). Between the ribbons, loops are observed in different wavelengths: bright X ray loops, in coronal lines 171 Å and 195 Å and bright or dark in Hα as they are cooling ([15], [16]). The loops form growing arcades with higher and higher hot loops while the reconnection point is rising. In the ribbons, kernels are the footpoints of loops ([17]). It was shown that the later arcades are less and less sheared ([18]). During cooling process the two ribbon distance is increasing. In complex active regions different systems of ribbons can be formed and developed in a fast time, of the order of few minutes (14 July 2000, 28 October 2003, 13 September 2005, 27 May 2003 - Figure 5). In the event of 13 September 2005 a series of 10 flares occur within 12 hours and even a series of 3 flares (X1.4, X 1.5, X1.7) in two hours. The concerned active regions are all complex containing a δ-spot with constant new emerging field with fast moving polarities (Figure 6). It is difficult to identify the trigger of each flare of the series.

Nevertheless, we are able to find the locations where the magnetic reconnection is possible and produce flares by analyzing the topology either by lfff (linear-force-free field) extrapolation when the shear is weak enough, otherwise by nlfff (non linear-force-free field) methods of extrapolation have to be applied in the latter case ([19]).

Many models of flare onset are developed based on tether-cutting or straining force ([20]). For large quadrupolar configuration, the breakout model proposed by [4]) seem to be successful for some cases ([21]). The existence of null point which was pointed out in the Antiochos model does not seem to be a necessary condition for breaking the overlying lines ([22], [13]). Reconnection could occur in quasi separatrices layers with no null points (Figure 7).

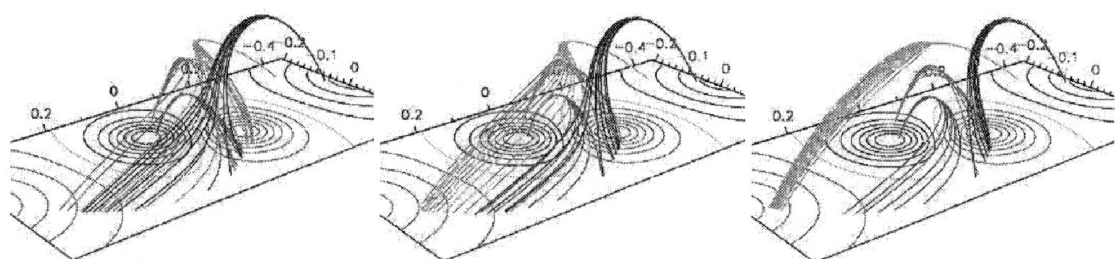

FIGURE 9. Quadrupolar magnetic reconnection with slip-running field lines for three different times. The darker field lines have a fixed footpoint anchored in a fixed point in the right source, the other footpoint is slip-running continuously along the quasi-separatrice layer. During the reconnection, the plasma in these different field lines will be successively heat with footpoints visible as bright kernels. This behaviour explains the fast extension of ribbons [24]

QSL RECONNECTION WITH SLIPPAGE OF MAGNETIC FIELD LINES

In the first section we have shown complex examples of flares with multiple ribbons formed at different places successively. Here we present a simple compact flare with no CME in order to study in details the dynamics of the ribbons, not the well known increasing distance of the ribbons versus time but the fast lateral extension of the ribbons. The flare occurs in a sheared active region at the location of a small emerging flux (Berlicki et al 2004). The observation in Hα shows two ribbons, one of them is relatively short and stays short, the other one is developing an extension in a few minutes (Figure 8). The conclusion of the [23] paper was that there was a series of reconnections in the corona. The question arises to know if this was really due to multiple reconnections. We performed a lfff extrapolation using the data base of FROMAGE and pointed out the existence of loop-like field lines growing with one fixed footpoint and the other footpoint was slip-running along the ribbon (Figure 9).

[24] propose a new idea with the slippage of the field lines along the QSLs. Using time dependent MHD simulation it is shown the possibility of field lines to change connectivity in a continuously way and not abruptly like in the classical null point reconnection where field lines clearly connect by pair. This process can take place in many solar flares. Magnetic field configurations with fields weakly stressed by asymmetric line-tied twisting motions are considered. When the line-tied driving is suppressed, magnetic reconnection is solely due to the self-pinching and dissipation of current layers formed along QSLs. For thin QSLs and high resistivity, the field line footpoints slip-run at super-Alfvénic speeds along the intersection of the QSLs with the chromosphere. Since particles are accelerated in the diffuse regions along the field faster than the Alfvén speed, slip-running reconnection could explain the fast extension of the ribbons in all the wavelengths, being the result of heat of the plasma as particles travel in denser plasma (Figure 9).

Recently Hinode observed moving loops moving that have been interpreted as due to the slippage of magnetic field lines ([25]).

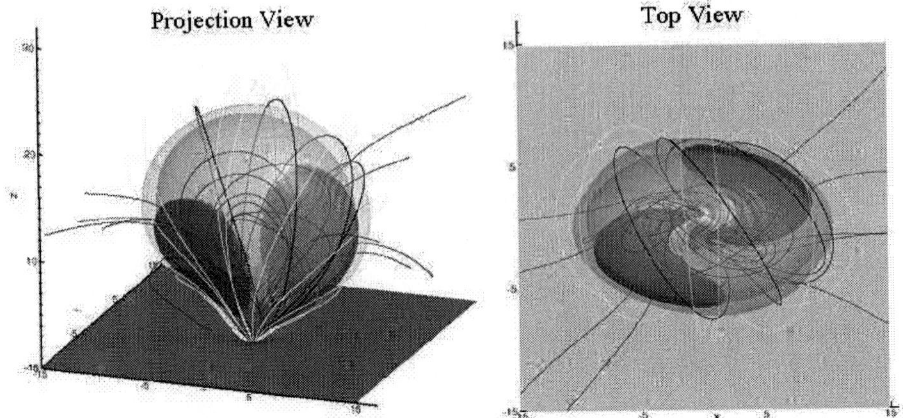

FIGURE 10. Numerically expanding flux tube using an MHD simulation viewed from above (right) and on side (left). These surfaces show the formation of the current shell between the expanding flux tube and the outer potential field. It simulates the EIT waves ([26].

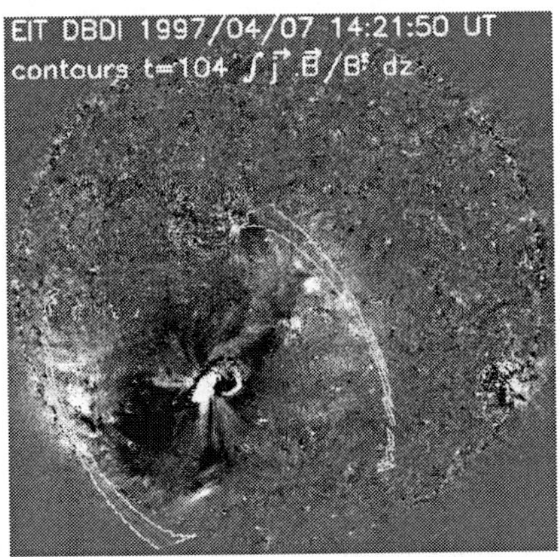

FIGURE 11. Overlay of observations of the active regions where the propagating EIT wave on April 7, 1997 took place with quantities resulting from the simulation. The EIT wave at 14:21:50 UT is overlaid by contours of the projected boundaries of the inflating separatrices. The EIT DBDI images are obtained from correcting the EIT images for the differential rotation of the Sun to the time of a pre-event image and subtracting the pre-event images afterwards. The time of the pre-event image is 14:00:03 UT on April 7, 1997.

EIT WAVES: ARE THEY REALLY WAVES?

EIT waves are observed in EUV as bright fronts propagating in the transition region (Figure 10). Some of these bright fronts propagate across the solar disc. EIT waves are all associated with a flare and a CME and are commonly interpreted as fast-mode magnetosonic waves ([27],[28],[29]). Propagating EIT waves could also be the direct signature of the gradual opening of magnetic field lines during a CME ([30], [31]). [26] and [32] quantitatively addressed this alternative interpretation. Using two independent 3D MHD codes, they performed numerical simulations of a slowly rotating magnetic bipole,which progressively result in the formation of a twisted magnetic flux tube and its fast expansion,as during a CME. They analyze the origins, the development and the observability in EUV of narrow electric currents layers which appear in the simulations. The results are confronted with two well-known SoHO/EIT observations of propagating EIT waves (April 7 and May 12,1997), by scaling the vertical magnetic field components of the simulated bipole to the line of sight magnetic field observed by SoHO/MDI and the sign of helicity to the orientation of the soft X-ray sigmoids observed by Yohkoh/SXT. A large-scale and narrow current shell appears around the twisted flux tube in the dynamic phase of its expansion (Figure 11). This current shell is formed by the return currents of the system, which separate the twisted flux tube from the surrounding fields. It also corresponds to a large-scale dynamically formed QSL. It intensifies as the flux tube accelerates and it is co-spatial with weak plasma compression. The current density integrated over the altitude has a shape of an ellipse which expands and rotates when viewed from above, reproducing the generic properties of propagating EIT waves. The timing, orientation and location of bright and faint patches observed in the two EIT waves are remarkably well reproduced (Figure 11). Propagating EIT waves can be interpreted as the observational signature of Joule heating in electric current shells,which separate expanding flux tubes from their surrounding fields during CMEs. [26] conjecture that the bright edges of halo CMEs show the plasma compression in these current shells.

CONCLUSION

Flares and eruptions are dynamical events which evolve with a fast speed. Past flare models were static in 2D and 3D. They provided good insight on the magnetic topology of active regions and the loci of possible deposit of energy.

We have shown that MHD simulations explain some particular aspects of the dynamics of flares. In one case we have shown how slip-running magnetic field lines in quasi-separatrice can mimic fast extension of Hα and UV ribbons obserrved by THEMIS and TRACE and running XRT loops observed by Hinode. In a second case simulations provide a new interpretation of EIT waves by the existence narrower current shells during the eruption of large flux tubes and the ambient magnetic field.

ACKNOWLEDGMENTS

B.S. wants to thank deeply the C. Dumitrache organizer of this workshop.

REFERENCES

1. T. G. Forbes, and P. A. Isenberg, *ApJ* **373**, 294–307 (1991).
2. J. Lin, T. G. Forbes, P. A. Isenberg, and P. Demoulin, *ApJ* **504**, 1006–+ (1998).

3. P. F. Chen, and K. Shibata, *ApJ* **545**, 524–531 (2000).
4. S. K. Antiochos, C. R. DeVore, and J. A. Klimchuk, *ApJ* **510**, 485–493 (1999).
5. P. Démoulin, "Magnetic Topologies: where Will Reconnection Occur ?," in *ESA SP-596: Chromospheric and Coronal Magnetic Fields*, edited by D. E. Innes, A. Lagg, and S. A. Solanki, 2005.
6. D. W. Longcope, *Living Reviews in Solar Physics* **2**, 1–58 (2005).
7. B. Schmieder, *Journal of Astrophysics and Astronomy* **27**, 139–149 (2006).
8. G. E. Vekstein, and E. R. Priest, *ApJ* **384**, 333–340 (1992).
9. P. Démoulin, L. G. Bagalá, C. H. Mandrini, J. C. Hénoux, and M. G. Rovira, *A&A* **325**, 305–317 (1997).
10. P. Démoulin, *Advances in Space Research* **37**, 1269–1282 (2006).
11. B. Schmieder, G. Aulanier, P. Démoulin, L. van Driel-Gesztelyi, T. Roudier, N. Nitta, and G. Cauzzi, *A&A* **325**, 1213–1225 (1997).
12. P. Démoulin, J. C. Hénoux, E. R. Priest, and C. H. Mandrini, *A&A* **308**, 643–655 (1996).
13. C. H. Mandrini, P. Demoulin, B. Schmieder, E. E. Deluca, E. Pariat, and W. Uddin, *Sol. Phys.* **238**, 293–312 (2006).
14. J. Lin, *Sol. Phys.* **222**, 115–136 (2004).
15. B. Schmieder, P. Heinzel, L. van Driel-Gesztelyi, and J. R. Lemen, *Sol. Phys.* **165**, 303–328 (1996).
16. G. D. Zanna, B. Schmieder, H. Mason, A. Berlicki, and S. Bradshaw, *Sol. Phys.* **239**, 173–191 (2006).
17. G. D. Zanna, A. Berlicki, B. Schmieder, and H. E. Mason, *Sol. Phys.* **234**, 95–113 (2006).
18. Y. N. Su, L. Golub, A. A. van Ballegooijen, and M. gros, *Sol. Phys.* **236**, 325–349 (2006).
19. S. Régnier, and R. C. Canfield, *A&A* **451**, 319–330 (2006).
20. J. Klimchuk, *Geographycal monograph* **125**, 143–157 (2001).
21. G. Aulanier, E. E. DeLuca, S. K. Antiochos, R. A. McMullen, and L. Golub, *ApJ* **540**, 1126–1142 (2000).
22. B. Schmieder, C. H. Mandrini, P. Démoulin, E. Pariat, A. Berlicki, and E. Deluca, *Advances in Space Research* **37**, 1313–1316 (2006).
23. A. Berlicki, B. Schmieder, N. Vilmer, G. Aulanier, and G. Del Zanna, *A&A* **423**, 1119–1131 (2004).
24. G. Aulanier, E. Pariat, P. Démoulin, and C. R. Devore, *Sol. Phys.* **238**, 347–376 (2006).
25. G. Aulanier, L. Golub, E. . DeLuca, J. W. Cirtain, R. Kano, L. L. Lundquist, N. Narukage, T. Sakao, and M. A. Weber, *Science* **1**, submitted (2007).
26. C. Delannée, J.-F. Hochedez, and G. Aulanier, *A&A* **465**, 603–612 (2007).
27. B. J. Thompson, J. B. Gurman, W. M. Neupert, J. S. Newmark, J.-P. Delaboudinière, O. C. St. Cyr, S. Stezelberger, K. P. Dere, R. A. Howard, and D. J. Michels, *ApJL* **517**, L151–L154 (1999).
28. Y. Uchida, *Sol. Phys.* **4**, 30–+ (1968).
29. S. T. Wu, H. Zheng, S. Wang, B. J. Thompson, S. P. Plunkett, X. P. Zhao, and M. Dryer, *JGR* **106**, 25089–25102 (2001).
30. C. Delannée, and G. Aulanier, *Sol. Phys.* **190**, 107–129 (1999).
31. P. F. Chen, C. Fang, and K. Shibata, *ApJ* **622**, 1202–1210 (2005).
32. C. Delannée, G. Aulanier, T. T., and J.-F. Hochedez, *Sol. Phys.* **1**, in press (2007).

Analysis of 3D Magnetic Reconnections of Two Active Regions

C. Oprea, M. Mierla and C. Dumitrache

Astronomical Institute of Romanian Academy, Bucharest, Romania, e-mail: consti@aira.astro.ro

Abstract. In this paper we analyze the 3D magnetic field configuration of two neighbor active regions observed in October 2001. These active regions are situated near a huge polar filament which had a complex dynamics. We investigate the connections and influences between these active regions and filament evolution. The AR09640 is a well developed region. In its neighborhood emerged the young AR09644 that produced few flares. Our study reveals the 3D magnetic reconnections in all the area.

Keywords: Solar activity, active regions, magnetic reconnections
PACS: 96.60.Iv, 96.60.Q-

INTRODUCTION

Magnetic reconnections are of particular importance in solar system plasma and they are thought to be at the heart of the energy conversion processes. Emergence of the photospheric magnetic flux has been considered the main cause for solar flares and even for coronal mass ejections (CMEs). Ground-based observations have also shown a strong relationship between flares and the emergence of new magnetic flux. The coronal magnetic field is anchored to the solar surface. An active region is an area on the Sun with an especially strong magnetic field. Its evolution manifests through all layers of the solar atmosphere and all the evolution is controlled by the magnetic field. Magnetic flux tubes emerge at the solar surface (photosphere) but often submerge, so the active regions rise and decay. Chromospheric and coronal active phenomena such as flares, ejections and general heating are often attributed to the manner in which the coronal field responds to photospheric motions ([1]). The transformation of magnetic energy takes various forms in the solar atmosphere, ranging from coronal mass ejections and flares, down to the very small events implied in quasi-continuous coronal heating ([2]).

In this paper we investigate the changes of coronal magnetic field lines extrapolated from the observations of the photospheric magnetic field provided by MDI/SOHO. Our attention was focused on the magnetic configurations of the 3D magnetic loops when flares or other phenomena occurred and the relationship between the AR09640 and AR09644 (situated in the Northern hemisphere) and a huge polar filament.

OBSERVATIONAL DATA AND METHOD

In order to analyze the active regions we have used the following data: magnetograms from MDI (Michelson Doppler Imager), EIT (Extreme-Ultraviolet Imaging Telescope)

CP934, *Flows, Boundaries, Interactions*
edited by C. Dumitrache, V. Mioc, and N. A. Popescu

195 Å movies from SOHO, 171 Å images from TRACE (The Transition Region and Coronal Explorer), $H\alpha$ images from BBSO (Big Bear Solar Observatory) and Meudon Observatory, and our own $H\alpha$ observations.

Using a dipolar magnetic field configuration, Jong Kwan - Jake - Lee developed a model to extrapolate the photospheric data and obtain the coronal magnetic loops ([3]). This author developed an IDL code to visualize the 3D coronal magnetic field. The positions of the dipole are obtained from the MDI magnetograms. The components of the magnetic field intensity (B_x, B_y, B_z) are computed as follows:

$$B_x = B_0 \frac{3xz}{r^5} \tag{1}$$

$$B_y = B_0 \frac{3yz}{r^5} \tag{2}$$

$$B_z = -B_0 \frac{(1 - 3\frac{z^2}{r^2})}{r^3}, \tag{3}$$

where x, y and z are the Cartesian components of the position vector for the field lines and r is the module of the position vector:

$$r = \sqrt{(x^2 + y^2 + z^2)}, \tag{4}$$

B_0 is the magnetic field intensity. Each magnetic field line was generated from a set of piecewise linear curves, with each curve obtained from a magnetic flux density vector $\vec{B}$. Each piecewise linear curve has a small length, ΔS (a piecewise linear curve is defined by the direction of the vector $\vec{B}$,with lengths ΔS, $\vec{B} \times \vec{\Delta S} = 0$).

Finding the locations of dipoles below the $z = 0$ plane starting from the B_z plots (the magnetogram) involves two steps. The first step is finding the xy coordinates of the major or dominate dipoles, and the second step is finding the z coordinates (the distance from the surface of the Sun to the dipole) of the dipoles. Since dipoles can be either positive or negative, the xy coordinates of the positive dipoles and the xy coordinates of the negative dipoles are found separately. The positive dipoles are located at the points of the magnetogram where B_z has its largest positive value and the negative dipoles are located at the points of the magnetogram where B_z has its most negative value. Therefore, the xy coordinates of positive dipoles can be found by using the local maxima of values and the xy coordinates of negative dipoles can be found by using the local minima of B_z values on the magnetogram.

The local maxima for the positive dipoles are obtained by comparing a B_z value to the B_z values of its neighbors. If a B_z value is larger than all of its neighbors B_z values, the xy coordinates of the B_z value are saved as the potential xy coordinates of a positive dipole. Only those points that have an associated B_z value that exceeds 20 percent of the maximum B_z value of the magnetogram are selected to be the xy coordinates of the positive dipoles. The xy coordinates of negative dipoles can be found using a similar method excepting this case, when the local minima of B_z values are used instead of the local maxima.

The location of the dipoles z coordinates from the magnetograms found by using a numerical method, such as Powell method by comparing the real magnetogram with an estimated magnetogram derived from a set of dipoles. This method is a technique that minimizes a function of two or more independent variables ([3];[4]).

THE ACTIVE REGIONS EVOLUTION

We have analyzed the active regions AR09640 and AR09644 in the period 30 September 2001 - 5 October 2001 (during Carrington Rotation 1981). These two active regions are situated at the Southern part of a huge polar filament. The AR09640 appears before 30 September, but has no major influence in the neighborhood, except the period after the appearance of AR09644 when it seems to interact with this. AR09644 appeared on the solar disk on 1 October, as a small β region and few days later it increased in intensity and became a compact β region.

On 2 October, the AR09644 displays a 3D magnetic field lines, detected by the method described above, as plotted in Fig. 1. In this figure we have added the images given by TRACE and MDI on this active region, in order to compare the accuracy of the method we have applied. The field lines plotted on the TRACE and MDI images are those from the TRACE web site, while at the bottom of the Fig. 1 is our visualization in 3D using the MDI data.

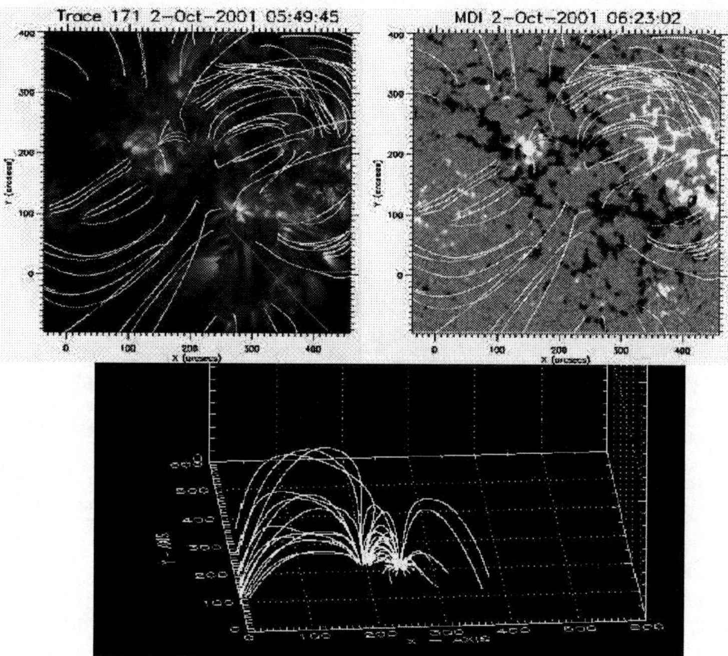

FIGURE 1. TRACE image and MDI magnetogram recorded on 2 October 2001, (upper panel); extrapolated magnetic field of the AR09644 (bottom panel).

On 4 October this region was a decaying β region connected with AR09640 via a set of magnetic loops. On 5 October AR09644 has almost totally decayed but it was still connected to AR09640 via magnetic loops; it is well known the fact that an active region disappears first in visible and later in magnetograms. AR09640 was a previously developed region, appeared on 27 September as a small β region. On 4 October it was a decaying β region, while on 5 October it was a decaying bi-polar plage.

Two events were linked to AR09644. First, a small $H\alpha$ flare reported on 3 October, having N18W26 coordinates. The flare had beginning at 14:30 UT, the maximum at 14:34 UT and end at 14:41 UT. The second event was a small X-ray flare occurred on 4 October, at N18W41, with the beginning at 14:16 UT, maximum at 14:19 and end at 14:26.

Using the method described in section 2 we analyzed the topology of the magnetic field of these two regions. We have observed strong magnetic reconnections in this area and tried to establish a connection between these and the succession of the events.

On 1 October AR09640 was a well developed region, while AR09644 emerged in its vicinity. Until this moment, no magnetic loops connected these two regions. We visualized magnetic loops linkage between them on 2 October, when AR09644 began to develop. Few hours later, magnetic reconnections could be observed and break out of the linkage between the AR09640 and AR09644 (fig.2).

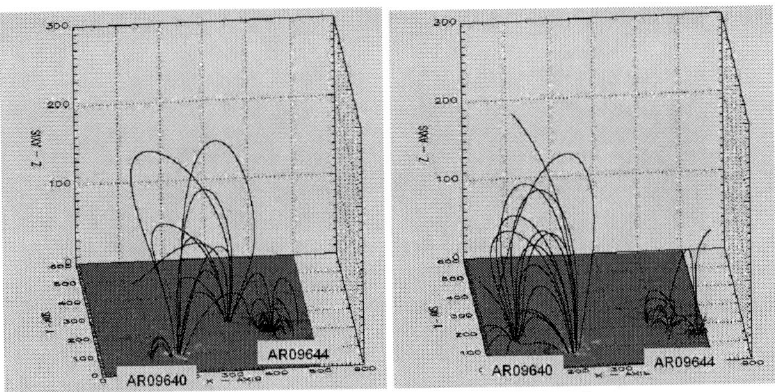

FIGURE 2. Extrapolated magnetic field linking AR09640 and AR09644 superimposed on MDI magnetogram, recorded on 2 October 2001, 07:00 UT (left panel); extrapolated magnetic field of the AR09640 and AR09644 superimposed on MDI magnetogram, recorded on 2 October 2001, 21:07 UT (right panel).

On 3 October, at 10:41 UT, before the first flare reported occurring in AR09644, the magnetic loops linking both regions were very strong (Fig. 3, left panel). In the same day, at 18:07 UT, after the flare produced, the magnetic field lines decreased in height and part of them opened into the corona (Fig.3, right panel).

Before the second flare, a small X-ray flare observed in AR09644, on 4 October, at 14:01 UT, both active regions were linked by magnetic loops. One can observe an increase of these loops in number and elevation, indicating a strong magnetic field in the zone (Fig. 4, left panel).

At 14:28 UT, after this X-ray flare produced, the magnetic field lines opened in AR09644 and disconnected from the AR09640. The AR09640 kept its topology and it seemed to be unchanged by the flare (Fig. 4, right panel).

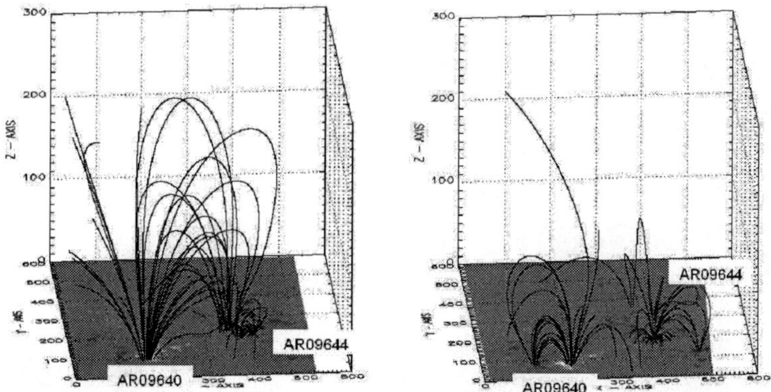

FIGURE 3. Extrapolated magnetic field of the AR09640 and AR09644 superimposed on MDI magnetogram, recorded on 3 October 2001, 10:41 UT (left panel); extrapolated magnetic field of the AR09640 and AR09644 superimposed on MDI magnetogram, recorded on 3 October 2001, 18:07 UT (right panel).

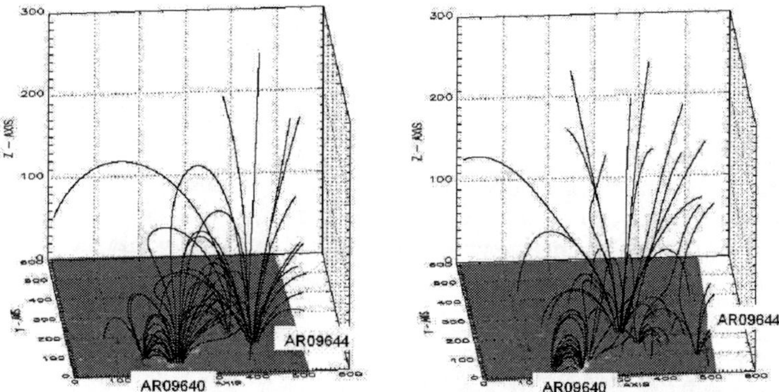

FIGURE 4. Extrapolated magnetic field of the AR09640 and AR09644 superimposed on MDI magnetogram, recorded on 4 October 2001, 14:01 UT (left panel); extrapolated magnetic field of the AR09640 and AR09644 superimposed on MDI magnetogram, recorded on 4 October 2001, 14:28 UT (right panel).

Later, magnetic loops closed and linked both regions, but AR09644 seemed to link also with other region in its neighborhood. At 23:00 UT, on 4 October, the magnetic field lines of AR09640 displayed open lines similarly to the topology observed after a flare, but no such event was reported in region. The magnetic field of the AR09644 decreased in intensity (Fig. 5, right panel) as indicates the elevation of the magnetic loops displayed on 3D by the described method.

In the last day of their existence, 5 October, both regions were connected again by magnetic loops and they were also linked to other active regions in their proximity. The magnetic loops decreased in height in AR09644. AR09640 was still linked to AR09644 but it had a stronger connection with another active region from neighborhood.

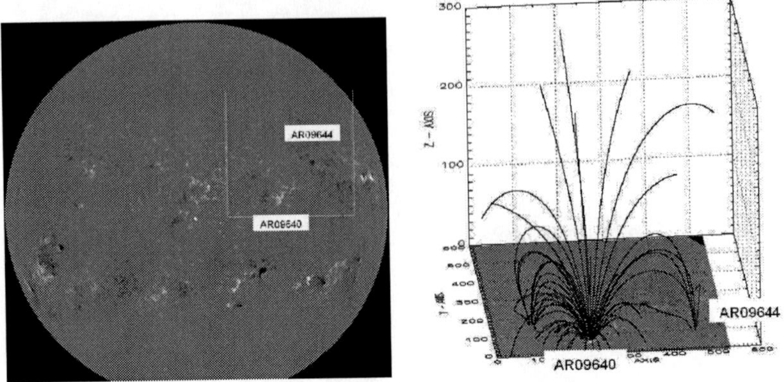

FIGURE 5. MDI magnetogram recorded on October, 4, 2001, 23:00 UT (left panel); extrapolated magnetic field of the AR09640 and AR09644 superimposed on MDI magnetogram (right panel).

In the same period, Northward of these two active regions, evolved a huge polar filament that had a complex dynamics, changing its shape, its thickness, its height and its orientation. Between 3 and 5 October a major change was observed in the filament evolution (day of year 276-278). On 5 October one observed a sudden variation of the tilt angle, after an increase of the differential rotation in the precedent days and a sudden decrease of it and quick variations during the next days. The bursting of the Western part of the filament produced on 5 October and the enlargement of this part produced on 6 October (Fig. 6). The filament length attended its minimum while the tilt angle and the computed magnetic field became maximum ([5]).

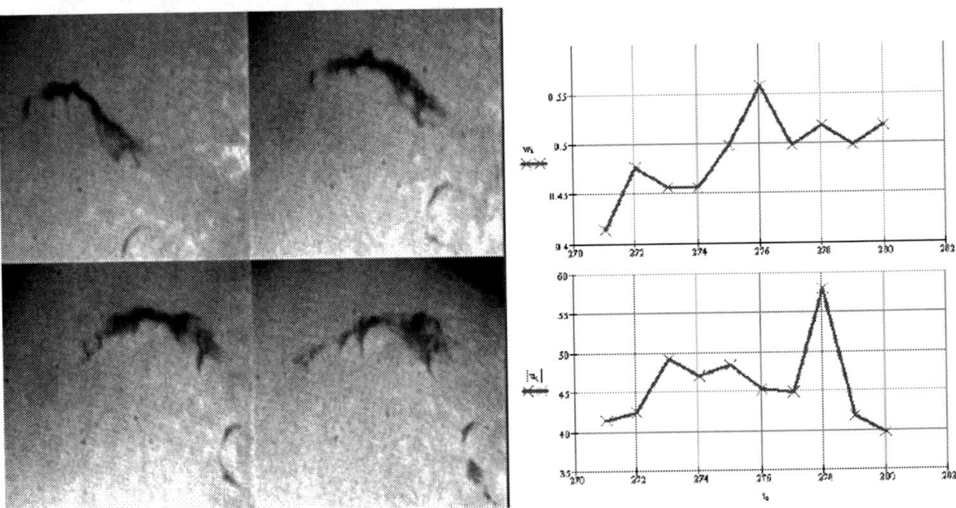

FIGURE 6. Filament observations at Bucharest Observatory on 3, 4, 5 and 6 October 2001 (left panel); the differential rotation (*w*) and the tilt angle (*u*) of the polar filament (right panel).

We think that the eruption observed on 4 October had a major influence on the filament evolution and its fragmentation. In these large scale magnetic reconnections, the AR09644 short evolution, with certain flares activity, played a major role. First, its emergence, its flaring activity and after that its evanescence - all these in few days produced important magnetic reconnections in the zone. These reconnections linked this small active region to its neighborhood more developed. The flares occurrence temporarily opened the magnetic field lines. All these large-scale reconnections influenced the filament evolution too. On 5 October, the filament changed suddenly its inclination on the solar parallel, after previous variations of its differential rotation. On 4 October, at 23:00 UT, the magnetic loops opened in AR09640. We suspect that important large scale magnetic reconnections produced this sudden change in the filament orientation and the opening of the loops of AR09640.

CONCLUSIONS

We have investigated the 3D topology of the magnetic field joining two active regions: one well developed region (AR09640) and one new emerged region (AR09644).

From the data we have analyzed here, we could see that:

- The young AR09644 produced two flares (a $H\alpha$ and a X-ray flare), facts revealed by the opening of the magnetic field lines and magnetic reconnections.
- These magnetic reconnections played an important role in a neighbor Northern filament dynamics.
- Our method of investigation revealed also an opening of the magnetic field lines in AR09640, where no flare observation was reported. The sudden changes of the Northern filament inclination and this lines opening could be a consequence of en energy release observable in other length that we have studied in this article.

ACKNOWLEDGEMENTS

We would like to thank the SOHO consortium for providing the data and the software libraries. SOHO is a project of international cooperation between ESA and NASA.

REFERENCES

1. S. Régnier, T. Amari, E. Kersalé, *A&A* **392**, 1119–1127 (2002).
2. P. Démoulin *Proceedings of the International Scientific Conference on Chromospheric and Coronal Magnetic Fields (ESA SP-596)*, 22.1–22.12 (2005).
3. J. K. Lee, "Coronal Loop Identification", *Master Thesis*, University of Alabama in Huntsville (2002).
4. W. H. Press, B. P. Flannery, S. A. Teukolsky and W. T. Vetterling, *Numerical Recipes in C: The Art of Scientific Computing*, 2nd Edition, Cambridge University Press, Cambridge (1993).
5. C. Dumitrache, C. Oprea, D. Constantin, M. Mierla, A. S. Popescu, in C. Dumitrache, N.A. Popescu, D.M. Şuran, V. Mioc (eds), *Fifty Years of Romanian Astrophysics*, *AIP Conf. Proc. Ser.*, Vol. **895**, 69–74 (2007).

Chromospheric Line Emission Analysis of the July 16, 2004 Sun Quake

Diana Beşliu-Ionescu[*,†], Alina Donea[*], Paul Cally[*] and Charles Lindsey[**]

[*]*CSPA, School of Mathematical Sciences, Monash University, Clayton, Australia, 3800*
[†]*Astronomical Institute of the Romanian Academy, Str. Cutitul de argint, Nr. 5, Bucharest, Romania, Ro-040557*
[**]*CoRA, 3380 Mitchell Lane, Boulder, CO, USA, 80301*

Abstract.
Observations in chromospheric lines and the visible continuum together with photospheric helioseismic measurements make possible to image a 3-dimensional profile of a sun quake in a flaring region. Chromospheric line observations show us the part of the solar atmosphere where high-energy electrons are thought to cause thick target heating that then could also cause intense white-light emission and could drive seismic waves into the active region subphotosphere. we present here the preliminary results of the sun quake of July 16, 2004.

Keywords: helioseismic holography, chromosphere, sun quakes, h alpha
PACS: 96.60.Ly

INTRODUCTION

Solar flares are one of the most interesting phenomena on the Sun. Such a powerful phenomenon can cause sun quakes. The first sun quake was discovered by Kosovichev & Zharkova [5]. Donea, Braun and Lindsey [6] subsequently imaged the seismic source of this quake. The quake manifested itself by producing acoustic waves that propagated thousands of kilometers from the location of the flare into the solar interior and back to the surface. Outgoing circular waves were observed 20 to 60 minutes after the impulsive phase of the flare, at distances from approximately 18 to 50 Mm from the flare site. Follow-up efforts to detect seismic emission from several other flares in 1998 and 1999, some considerably larger than the flare of July 9, 1996, showed no other instances of significant acoustic emission.

Helioseismic holography (Lindsey and Braun [9]) is a technique used to probe compact structures on and below the solar surface, such as active regions and sunspots. It has been used to detect enhanced acoustic emission from more than a dozen solar flares (Donea, Braun and Lindsey [6], Donea & Lindsey [7], Beşliu-Ionescu et. al [2]).

We have studied the seismic (Donea & Lindsey [7]; Beşliu et. al [1], Beşliu-Ionescu et. al [3]) and magnetic properties (Beşliu et. al [1]) of active regions in order to understand the acoustics of solar flares.

CP934, *Flows, Boundaries, Interactions*
edited by C. Dumitrache, V. Mioc, and N. A. Popescu

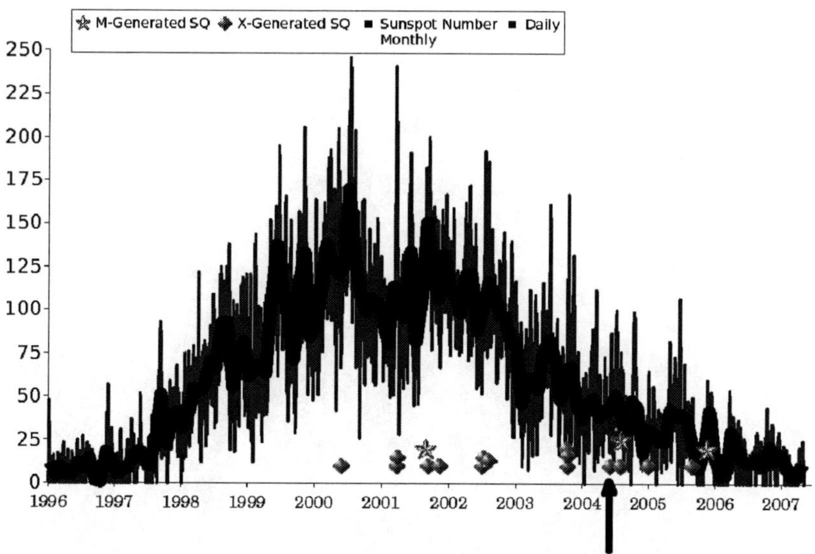

FIGURE 1. Acoustically noisy flares by comparison with the solar cycle 23. The arrow shows the position in the solar cycle 23 of the acoustically active flare of July 16, 2004.

DATA ANALYSIS

After extensive studies of seismic properties (Beşliu et. al [1], Moradi et al. [10], Donea et al. [8]) of these flares it was clear that a study of the chromosphere is a very interesting step forward towards understanding sun quakes morphology. We present here a detailed analysis of the Hα center line and its red wing for the acoustically noisy flare of July 16, 2004. The images were taken from the Improved Solar Observatory Optical Network (Neidig et al. [ISOON, see 11]) and have a 0.1 Å bandwidth centered on 6562.808 Å which we have focused on AR 10649 (-500",-200") with a field of view of 160×160 arcsec. In three days this active region produced seven X-type solar flares, but only one of these was seismically active: July 16, 2004, X3.6-type, which started at 13:49 UT, peaked at 13:55 UT and ended at 14:01 UT. It had an 3B Hα flare type associated with.

Figure 1 shows the seismically active flares known to date, against the monthly (thick line) and daily (thin line) sunspot number. The diamonds and stars represent sun quakes generated by X-type and respectively, by M-type solar flares. They were discovered during our survey of solar flares (Beşliu-Ionescu et. al [2], Beşliu-Ionescu et. al [4]) using SOHO/MDI Dopplergram data with good coverage (at least 2 continuous hours of observation, with one minute cadence). Taking into consideration that SOHO/MDI's data did not cover most of the flares from 1997 to 2000, we cannot positively say something about their occurrence with respect to this solar cycle. The sun quake of the July 16, 2004 solar flare is identified in this figure by the thick arrow.

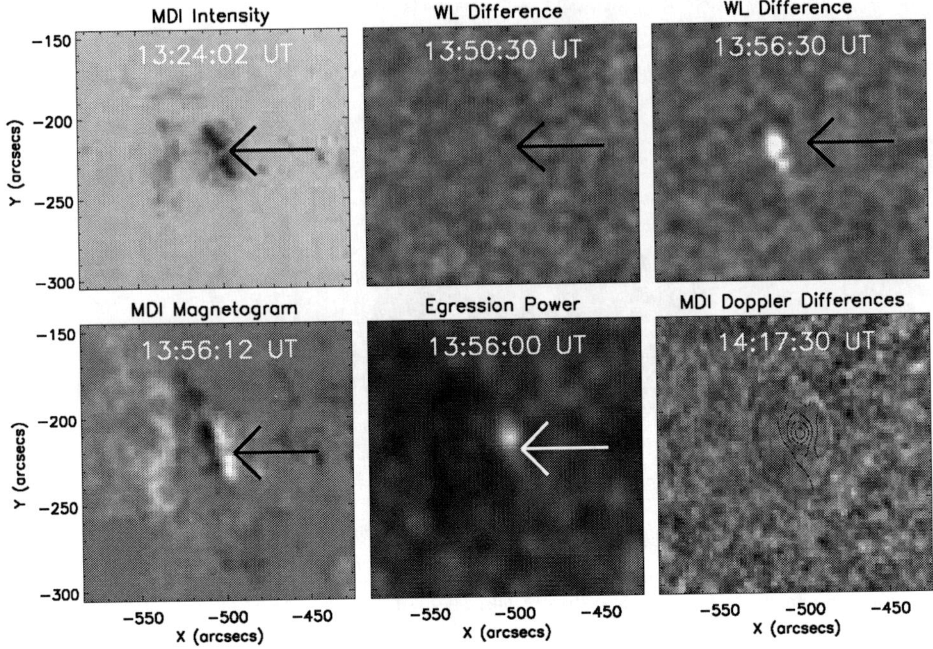

FIGURE 2. General presentation of the properties of AR 10649.

Figure 2 presents a general view of this active region. The arrow in all panels is pointing to the egression power signature (the seismic source produced by the flare), seen in the middle panel of the second row. First row left panel shows an intensity continuum image of this region. The next two panels show intensity continuum differences five minutes before and at the maximum of the sun quake. The white light signature is very clear and fairly consistent as structure and position with the acoustic signature. The first frame on the second row is a SOHO/MDI magnetogram. The seismic emission is co-aligned with the neutral line of the magnetic field. The middle frame is a smeared map of the 6 mHz egression power at the maximum of the sun quake. The last frame on the second row shows seismic waves expanding from the source about 20 minutes later. Superimposed are the contour levels of the maximum seismic emission (30%, 50%, 70% and 90%).

The Hα one minute cadence data that we have studied begin 30 minutes before the start of the flare and end 45 minutes later than the end of the flare. Figure 3 shows the evolution of the Hα emission in time.

One of the most important and interesting characteristics of the Hα emission is the central Hα kernel which has the same orientation (north-south) as the acoustic emission seen in the middle panel on the second row Figure 2. There is a strong spatial correlation between the Hα bright kernel see arrow in Figure 3 and the seismic observations of the acoustic source. The seismic signature also has a main kernel accompanied by a more diffuse seismic emission at 6 mHz.

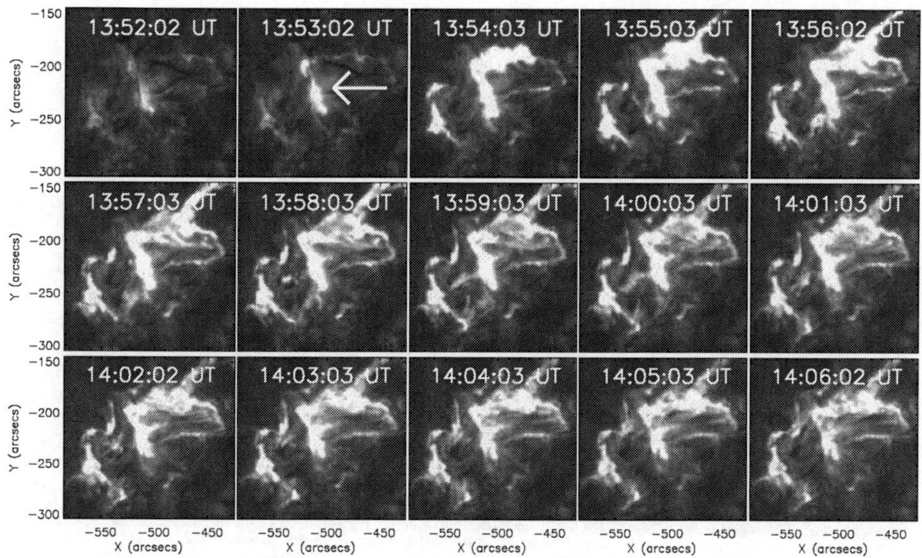

FIGURE 3. The evolution of AR 10649 in the Hα line center during the seismic emission. Each frame has attached its corresponding time. The arrow shows the location of the seismic source.

The temporal evolution of this feature is better seen in Figure 4 which is a sequence of Hα red-wing differences at the times specified in each frame. The specific times have been calculated as the middle time interval between the two differentiated frames. The strong variation in the third frame of Figure 4, shows a rapid downward movement of the chromospheric plasma which developed approximately three minutes before the maximum of the seismic source. The Hα kernel brightening had its maximum at 13:55 UT, just a minute before the maximum of the egression power seismic source.

INTERPRETATION AND DISCUSSION

We presented here a preliminary analysis of the Hα chromospheric line center for the seismically active flare of July 16, 2004. Hα emission presents individual flaring kernels of which the most bright one spatially coincides with the seismic source (whithin alignment uncertainties).

Most Hα emission from flares is the result of excitation of neutral hydrogen by a large population of coronal electrons The impact of the high-energy electrons, in the impulsive flare phase gives rise to strongly red-shifted chromospheric emission in a much more compact region, and this occurs at or near the site of seismic emission. A similar example, but for the Na D_1 line is shown in the middle right panel of Figure 8 in Donea & Lindsey [7].

Using Hα images we can investigate the precipitation sites of the particles. Energetic

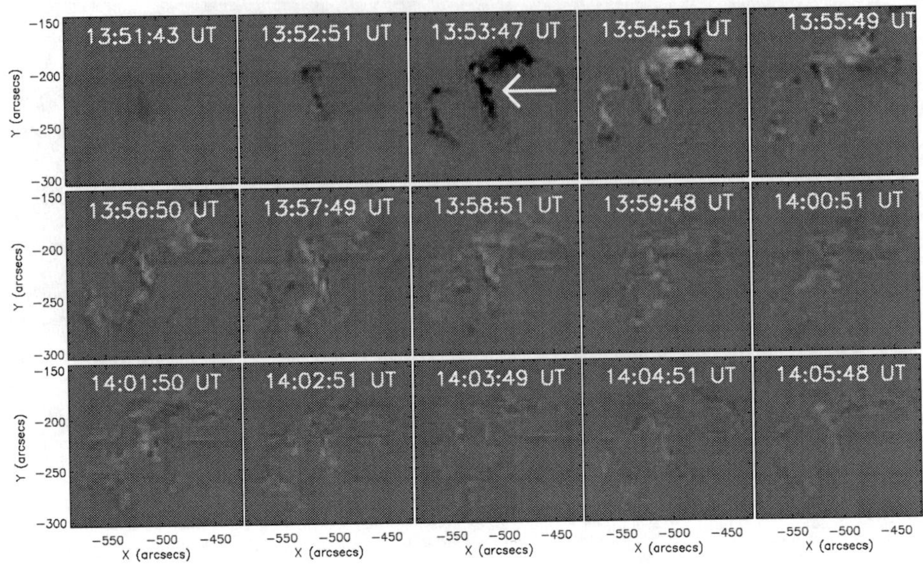

FIGURE 4. The evolution of AR 10649 in Hα red wing during the seismic emission shown as differences between two consecutive minutes. Each frame has its corresponding time on top, calculated as the middle of the time interval between the two frames which were differentiated. The arrow shows the location of the seismic source

electrons are being stopped here, in the chromosphere. The flare energy is transported from the reconnection site into the chromosphere, and further into the photosphere, possibly by "back-warming" mechanism (Donea et al. [8]), therefore igniting the sun quake.

A time resolution analysis of the X-ray data would have been needed for a complex analysis. Unfortunately RHESSI did not observe this flare at that time.

Hinode is expected to obtain high-resolution, high-cadence Hα and Na D-line observations of flares in the Solar Cycle 24, seismic observations of which will be available from *SDO*/HMI. These observations will be crucial for a realistic assessment of the contribution of chromospheric shocks to flare seismic emission into active region sub-photospheres.

REFERENCES

1. D. Beşliu, A.-C. Donea, P. Cally and C. Lindsey, *Rom. Astron. J.*, Bucharest, **15**, 17-32 (2005).
2. D. Beşliu-Ionescu, A.-C. Donea, P. Cally and C. Lindsey, *The Dynamic Sun: Challenges for Theory and Observations*, Edited by D. Danesy, S. Poedts, A. De Groof and J. Andries. Proceedings of the 11th European Solar Physics Meeting, Published on CDROM., 2005, p.111.1.
3. D. Beşliu-Ionescu, A. Donea, P. Cally and C. Lindsey, *Rom. Astron. J.* Supplement, Bucharest, **16**, 203-206 (2005).
4. D. Beşliu-Ionescu, A. Donea, P. Cally and C. Lindsey, *Proceedings of SOHO 18/GONG 2006/HELAS I,*

Beyond the spherical Sun (ESA SP-624) edited by Karen Fletcher. Scientific Editor: Michael Thompson, Published on CDROM, 2006, p.67.1

5. A. Kosovichev and V. Zharkova, *Nature*, **393**, 317-318 (1998).
6. A.-C. Donea, D. Braun and C. Lindsey, *ApJ* **513**, L143-L146 (1999).
7. A.-C. Donea and C. Lindsey, *ApJ* **630**, 1168-1183 (2005).
8. A.-C. Donea, D. Beşliu-Ionescu, P. Cally, C. Lindsey and V.V. Zharkova, *Solar Physics*, **239**, 113-135 (2006).
9. C. Lindsey and D. Braun, *ApJ*, **485**, p.895 (1997).
10. H. Moradi, A.-C. Donea, C. Lindsey, D. Beşliu-Ionescu and P. Cally, *MNRAS* **374**, 1155-1163 (2007).
11. Neidig, D. Wiborg, P., Confer, M. et al., The USAF Improved Solar Observing Optical Network (ISOON) and its Impact on Solar Synoptic Data Bases, Ed. K. S. Balasubramaniam; Jack Harvey; and D. Rabin, Synoptic Solar Physics – 18th NSO/Sacramento Peak Summer Workshop held at Sunspot; New Mexico 8-12 September 1997. ASP Conference Series; Vol. 140; p.519, 1998.

Sampling the Coronal Magnetic Field

Istvan Ballai

Solar Physics and Space Plasma Research Centre (SP^{2}RC), Dept. of Applied Mathematics, University of Sheffield, Hounsfield Road, Hicks Building, Sheffield, S3 7RH, United Kingdom

Abstract. High resolution observations of the last decade made it clear that the magnetic field is, without doubt, one of the most important quantities of the solar atmosphere. Magnetic field controls the dynamics, stability and thermal state of the plasma. At the same time, magnetic field is connecting the solar interior to the atmosphere and all dynamic events in the upper layers of the Sun have their instigator on the solar surface or the region underneath. In this contribution I will review some properties of magnetic field generation and expansion into the solar atmosphere. Some direct and indirect diagnostic techniques will be discussed with special emphasis on indirect methods which involve the properties of waves and oscillations used to give values for the magnetic field in the solar coronal structures and quiet Sun.

Keywords: Coronal magnetic field, field diagnostics, coronal seismology
PACS: 96.60.P-, 96.60.Hv, 96.50.Tf

INTRODUCTION

The high-resolution ground- and space based observations of the last decade have changed our view on the solar magnetism in many ways.

Many of the basic physical parameters of the Sun's outer atmosphere, the corona, cannot be measured directly (e.g. magnetic field, transport coefficients). The solar atmosphere, however, is dominated by magnetism therefore the determination of coronal magnetic field, the parameter which controls the dynamics, the structure and evolution of the corona of the Sun, has been a challenge for solar physicists for several decades.

The intensities of magnetic fields in the solar atmosphere are well above that of the Earth (e.g., in the sunspots the magnetic field is about 3000 G while on Earth this is about 1 G). The quantitative importance of the magnetic field is given by the plasma beta (the ratio of the thermal and magnetic pressure). While in the solar interior and low atmosphere this quantity is well above 1, starting from chromosphere up in the corona and beyond, the magnetic pressure dominates over the gas pressure meaning that magnetic field takes over the control of the plasma motion and stability.

Multi-satellite (Skylab, SOHO, TRACE, YOKHO) and multi-wavelength (white-light, Hα, (E)UV, X-ray) observations in the solar atmosphere showed that the locations of field concentration down in the photosphere correspond to regions of high emissivity of the solar material, i.e. large temperatures, in the solar corona. These sort of observations also revealed that even the heating of the solar corona has a magnetic nature. Time dependent (AC) and independent (DC) (compared to the Alfvén transit time) stresses are produced in the convective zone, propagate channelled by magnetic field lines and are dissipated effectively in the upper chromosphere and corona.

The difficulty studying generation and evolution of the magnetic field resides in its

CP934, *Flows, Boundaries, Interactions*
edited by C. Dumitrache, V. Mioc, and N. A. Popescu

very own complexity. Characteristic scales of magnetic structures vary between very large entities, clearly visible in different wavelengths (e.g. sunspots, coronal loops, plume/interplume, prominences) to very tiny filamentary structures which cannot be resolved using the present resolution. However, theoretical and numerical methods were - and are - developed in order to describe phenomena related to the solar magnetic field.

Obviously we cannot cover all likely diagnostic methods of magnetic field, nor are aiming for a complete description of the manifestation of solar magnetic field, therefore our selection may be biassed in spite of our best intention.

GENERATION, EMERGENCE AND EVOLUTION OF THE MAGNETIC FIELD

Today it is widely accepted that electric currents flowing in the solar plasma generate magnetic field. The mechanism responsible for this generation is the magnetohydrodynamic (MHD) dynamo working near the bottom of the convective zone, below the photosphere (about 200 Mm below the solar surface). Magnetic field starts to be fragmented and structured even at this depth. Helioseismic studies and measurements have shown that a strong radial shear in the rotation of the solar interior between the convective envelop and the rigidly rotating solar interior ensures that the solar dynamo is a viable generating mechanism (see, e.g. [1], [2]). At the base of the convective zone, the toroidal magnetic field is around 1.5×10^5 G ([3]), somehow larger that the value expected from the equipartition of magnetic energy with convective energy. This strength of the field is also required to satisfy the observational constraint of the tilt angle, i.e. the angle between the connecting line of the two polarities of an active region and the equator. The magnetic field lines accumulate in magnetic ropes or magnetic flux tubes. These flux tubes become unstable to buoyancy and, as a result, they rise towards the solar surface. During their move through the solar convective zone, the field strength decreases rapidly due to decrease in the gas pressure. At the same time, due to the turbulent motion in the convective zone, the magnetic flux tube start to fragment into smaller flux tubes. This fragmentation, higher in the upper solar atmosphere, gives rise to coronal magnetic loops with different sizes.

After their emergence magnetic flux tubes form bipolar active regions at the surface (in form of, e.g. sunspots). The duration of the rise of the magnetic tubes is of the order of a few months, the rising time being inversely proportional with the initial field strength. Since the Coriolis force tends to suppress motions perpendicular to the axis of rotation, magnetic flux tubes are forced, during their rise, to move and emerge at higher latitudes. This migration of magnetic tubes can be balanced by other forces (e.g. buoyancy) which could dominate over rotational forces ([4]). The number active regions emerging per unit time on the solar surface increases strongly with decreasing total flux of the region. The mixed polarity field seen everywhere on the quiet Sun on the supergranule scale is called the magnetic carpet ([5], [6]). Similarly to granulation, the photospheric and atmospheric magnetic field is also very dynamic. The structure of this small-scale field changes rapidly on very short spatial and time scales and flux continuously emerges and disappears nearly homogeneously over the surface. Interestingly, the smallest magnetic structures show no correlation with the solar cycle, therefore they are believed to originate in a separate, small-scale dynamo process.

By the time the magnetic flux tubes reach the photosphere they are strongly filamented with typical magnetic strengths of about 1.5kG, which is, again, larger than the equipartition field strength. The occurrence of magnetic field at this level is rather sporadic, it is believed that it covers only about 1% of the solar surface. Beside the bipolar active regions, a typical magnetogram of the solar surface (see Figure 1) also shows large areas of apparently unipolar field in a network pattern, which outlines the boundaries of large convective cells, in particular large unipolar regions could be always found around the poles, suggesting the existence of a global dipole component of the solar magnetic field, which is nearly aligned to the rotation axis.

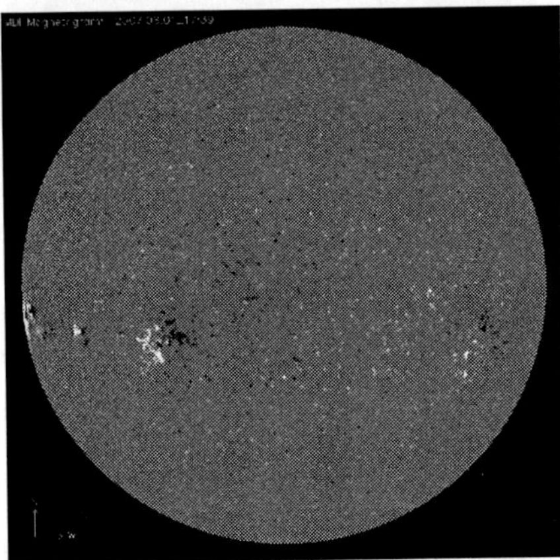

FIGURE 1. A typical magnetogram of the quiet Sun. The white and black features show a strong bipolar structure. The field is seen to be mixed polarity within and on the boundaries of the supergranule network. The figure has been adapted from the MDI magnetogram list available at the URL address *http : //soi.stanford.edu/production/*

Current data show that except in the vicinity of sunspots and pores the magnetic field is contained in the intergranular lanes. Observations show that the magnetic field outside active regions is very dynamic, continuously rearranging itself due to external convective agents. Measurements show that the strength of the magnetic field in these locations is about 1.2 kG and the magnetic field is concentrated in the 100 km wide core of these lanes. Because the convective cells evolve on a life timescale of about 5 minutes, the small-scale magnetic field pattern must also evolve on a similar time-scale. The edges of granules have a typical velocity of 6-10 km s^{-1} and may exceed the sound speed (10 km s^{-1}). As a result, the magnetic flux elements at the cell edges are buffeted on a time scale of 10-50 s. The quiet Sun magnetic field is replaced in approximately 10-14 hours.

Spectroscopic observations in circularly polarized light, which originates from the magnetic photosphere can show details of the internal structure of magnetic flux tubes. These results indicate that the photospheric layers of small flux tubes are hotter than their environment and that the plasma within the tube is in a continuous motion driven by flows, waves, and shocks. The magnetic tubes are surrounded by a downflowing gas and often show rapid horizontal displacement. This dynamical property of the magnetic flux tubes is relevant for channelling kinetic and magnetic energy from the photosphere to the solar upper regions where it contributes to the heating of the chromosphere and corona (see, e.g. [7]).

As we move higher up, the gas pressure decreases, therefore, magnetic flux tubes expand, and even merge with their neighbors. During their horizontal and vertical expansion magnetic field starts filling-up the entire available space. The almost horizontal field in the chromosphere is called magnetic canopy ([8], [9]). Above this height, the plasma becomes magnetically dominant, i.e. the magnetic pressure is larger than the gas pressure. The importance of the magnetic field in the control of the dynamics, stability and thermal state will persist until far in the solar wind. Above the canopy, the field becomes fairly homogeneous in strength with height but becomes inhomogeneous in directionality. In the higher regions of the solar atmosphere, magnetic flux elements show an average inclination of about 35-45 degrees (see, e.g. [10]).

Since the magnetic Lorentz force has no field aligned component, magnetic field serves as a guide for any disturbances and material flows. The collision rate between particles decreases with height, and therefore a strong anisotropy will develop with heat and particles being transported easier along the field rather than perpendicular to it. In the corona the orientation of the magnetic field can take any value. The topology of the magnetic field in the solar corona can be divided into closed regions (e.g. coronal loops connecting regions of opposite polarity) and large open unipolar regions with mainly vertical (radial) magnetic field. The differences in these two type of structures is apparent in their observation. While closed structures are bright in in X-ray and EUV emissions, the open regions seem more darker since the density in these structure is strongly diminished. Since the plasma is nearly frozen into the plasma, closed structures have no (or very little) material escape into the solar wind. In contrast, observations show that a vast amount of energy (about 10^6 erg cm^{-2}s^{-1}), momentum and material is lost via the open structures, such as coronal holes.

Depending on the location of the footpoints of magnetic field lines adjacent field lines need not be parallel to each other. These sort of field lines can form tangential discontinuities, or current sheets (see, e.g. [11]) which may play an important role in the process of coronal heating. When field lines cross each other, the topology of magnetic field lines changes, field lines reconnect and the entire structure evolve into another energetically favorable stage. During reconnection, a very intense DC current sheet if formed which can be dissipated to the surrounding plasma via, e.g. Joule heating. Every reconnection is accompanied by impulsively ejected shocks waves, energized particles, high-speed jets and radiation. This scenario has been put forward to explain the mechanism behind many energetic phenomena such as flares (on all observed energy scales) and CMEs (see, e.g. [12], [13], [14]).

Flares and CMEs are also responsible for generating strong blast waves (similar to the blast wave of an atomic bomb or the blast generated as a result of a supernova explosion) able to propagate across the extended solar disk. Behind the shock front, an extended density dimming is formed which is visible in base difference images. In the low pressure region behind the blast magnetoacoustic waves can propagate across the disk (usually these periodic or quasi-periodic disturbances are called EIT waves since first they were recorded using the SOHO/EIT instrument, e.g. [15]). At the same time these disturbances can interact with other magnetic entities in the solar corona (coronal loops, prominences, coronal holes, active regions) being absorbed, refracted or reflected. One of the most spectacular evidences of the interaction between blast waves and coronal loops are the oscillations generated in loops as a result of the collision. The global coronal EIT wave and the generated loop oscillations form the phenomenological basis of coronal magneto-seismology, one of the newest diagnostic branches of modern solar physics. The next section will be partly devoted to the presentation of the idea of coronal seismology, the methods used in evaluating plasma and field parameters as well to some results obtained by means of this method.

MEASUREMENT OF THE CORONAL MAGNETIC FIELD

In this section I will review some of the direct and indirect methods used to measure the coronal magnetic field, with special emphasis on the indirect method based on the diagnostic facility offered by MHD waves. For an excellent review on some of the techniques used to infer values for magnetic field see, e.g. [16]

Direct methods

Usually, direct measurement of the magnetic field is based on the Zeeman and Hanle effect, both being widely used with high precision in the solar photosphere. In the higher atmospheric layers the measurement of the magnetic field is a rather difficult task. Since the temperature rises from a few thousands to a few million degrees, atoms in the coronal plasma become highly ionized and most of their spectral lines shift into EUV range where the Zeeman splitting is very small while the lines are thermally broaden due to the high temperature. Therefore direct measurement of splitting of spectral lines due to magnetic field using Zeeman effect is impossible. However, in infrared wavelength, although the splitting is still not large enough to be resolved, it is possible to infer the line of sight component of the magnetic field ([17])

Other possibility is the use of the Hanle effect - which consists in the modification of the degree and orientation of the linear polarization in the presence of a magnetic field - but the precision of the measurement and its spatial resolution is much below the magnetic field measurement in the photosphere. In the solar prominences, Hanle effect gives the three component of the field (including electron density) from the polarization measurements of two spectral lines (the 10747Å line of Fe XIII and 5303Å line of Fe XIV) with different optical thicknesses ([18], [19]).

The coronal magnetic field is reconstructed (extrapolated) from the photospheric magnetic field observations (line of sight or vector magnetograms) assuming a force-free field (not exactly true for photosphere but applicable for corona) where the photospheric magnetograms serve as boundary conditions.

Radio measurement of coronal magnetic field has become a very popular diagnostic tool (see, e.g. [20], [21]). This technique exploits the fact that the electron gyrofrequency for typical coronal magnetic field lines in active regions lies in the radio frequency domain. Accelerated particles rotating around magnetic field lines are making the coronal material optically thick and these kind of regions will be highlighted in radio images. In order to describe the full 3D structure of the magnetic field in the corona extrapolations are still needed since the radio measurements do not measure the height of the gyroresonance layers (in a similar way as optically thin measurements cannot predict the location of the source in the line of sight). Using this technique relative strong magnetic field strengths (above 300 G) can be measured with high precision in 3D. Despite the accuracy of the method, radio measurements cannot diagnose weak field in the quiet Sun or coronal holes.

The magnetic field strength in the high corona, at altitudes between 2 and 9 $R_\odot$ above the photosphere can be measured through the Faraday rotation of a linearly polarized signal from interplanetary spacecraft or from cosmic radio sources in its way through the solar corona ([22]).

Indirect methods: Magneto-seismology

High resolution observations made in the last decade have changed dramatically our perception about the dynamical phenomena occurring in different layers of the solar atmosphere. Waves, which until mid 90s were sporadically observed (mainly in radio frequency), have become now a reality, waves being observed in all sort of wavelengths. The new achievement also meant that new areas of solar physics can have now an observational background. One of topics of solar physics which undergone a fascinating development is magneto-seismology, in particular *coronal seismology* ([23], [24], [25], [26], [27], [28]) which has been developed with the aim of serving an additional indirect tool for magnetic field diagnostics. The physical background of this theory resides in the fact that the observed waves are in the domain of frequencies corresponding to the MHD limit, i.e. all oscillations are somehow affected by the presence of the magnetic field.

Apart from the magnetic field, coronal seismology has proved to be a viable technique to sample the density structure in coronal loops, transport coefficients, heating function, density scale-height, and even predicting the sub-structuring of the magnetic field. Coronal seismology is using the observable properties of waves (amplitude, period, decay time and length, signature) combined with theoretical predictions on wave evolution (dispersion relation, damping lengths and times, etc.) to determine remotely plasma and field parameters in the corona. Waves have been observed to propagate in various magnetic structures such as sunspots, prominence fibrils, coronal loops, plume/interplume, etc. ([29], [30], [31], [32], [33]). Since these waves and oscillations are strictly localised in the structure in which they propagate they are labelled as *local waves*. Observations show that the observed waves have a very strong damping, practically they disappear after a few periods. This strong damping makes the subject of recent studies where the most likely explanation could be the damping due to resonant absorption, a mechanism which is possible in a plasma with transversal inhomogeneity ([34], [35],[36], [37]).

One of the first coronal seismology studies was carried out by [25]. First they studied the decaying oscillations in a coronal loop and they identified these oscillations as fast kink waves whose dispersion relation was studied and derived earlier by [38] in the form

$$\rho_e(\omega^2 - k^2 v_{Ae}^2) m_0 \frac{I_1'(m_0 R)}{I_1(m_0 R)} + \rho_0(\omega^2 - k^2 v_{A0}^2) m_e \frac{K_1'(m_e R)}{K_1(m_e R)} = 0, \quad (1)$$

where ρ_0, ρ_e are the densities inside (index "0") and outside (index "e") the loop, R is the radius of the loop, ω is the frequency of oscillations, k the longitudinal wavenumber, v_{A0}, v_{Ae} are the Alfvén speeds inside and outside the coronal loop, and the functions $I_1(x)$, $K_1(x)$ are modified Bessel function of the first order with the prime denoting the derivative with respect to the argument. The parameters m_0 and m_e, which can be considered as radial wavenumbers of the perturbations inside and outside the tube are defined as

$$m_i = \frac{(k^2c_{Si}^2 - \omega^2)(k^2v_{Ai}^2 - \omega^2)}{(c_{Si}^2 + v_{Ai}^2)(k^2c_{Ti}^2 - \omega^2)}$$

where $i = 0, e$ and the quantities c_S and c_T are the propagation speed of sound and slow waves, respectively. The period of these kink waves is given by

$$P_{kink} = \frac{2L}{c_K}, \tag{2}$$

where L is the length of the coronal loop and c_K is the propagation speed of kink oscillations given by

$$c_K \approx v_{A0}\left(\frac{2}{1+\rho_e/\rho_0}\right)^{1/2}$$

Based on these estimations [25] obtained that the magnetic field in the loop they studied by 13 ± 9 G, a value with an enormous error bar due to the lack of observation accuracy. Since then, all sort of coronal waves and oscillations have been used (slow and fast sausage and kink modes) each helping to derive valuable plasma and field quantities with increasing precision ([39]).

A different approach is needed when discussing the seismological possibilities of *global waves* which originate from impulsive and/or eruptive sources such as flares or coronal mass ejections (CMEs) and are able to travel over very long distances, sometimes comparable to the solar radius. While local coronal seismology tries to give answers on the strength and structure of the magnetic field in structures like coronal loops, global seismology provides estimates for magnetic field in the quiet Sun. The problem of coronal seismology and coronal loop oscillation should be seen as a global picture when the side-way action of a coronal global wave interacts with a coronal loop triggering local waves and oscillations.

Unambiguous evidence for large-scale coronal impulses initiated during the early stage of a flare and/or CME has been provided by Extreme-ultraviolet Imaging Telescope (EIT) observations onboard SOHO and by TRACE/EUV. EIT waves propagate in the quiet Sun with speeds of 250-400 km s^{-1} at an almost constant altitude. EIT waves have two stages: first there is an early (driven) stage where the wave correlates to radio type II bursts due to the creation of an energized population which serves as the source of radio emission. The second stage consists of a freely propagating wavefront which is observed to interact with coronal loops (see, e.g. [40]). Using TRACE/EUV 195 Å observations, [41] have shown that EIT waves (seen in this wavelength) are waves with average periods of the order of 400 s. Since at this height, the magnetic field can be considered vertical, EIT waves were considered fast MHD waves.

For an average value of EIT wave speed of 300 km s^{-1} propagating at 0.05 $R_\odot$ above the photosphere we find that the magnetic field is 1.8 G. If we apply $Br^2 = const.$, *i.e.* the magnetic flux is constant, we find that at the photospheric level the average magnetic field is 2.1 G which agrees well with the observed solar mean magnetic field ([42]) EIT

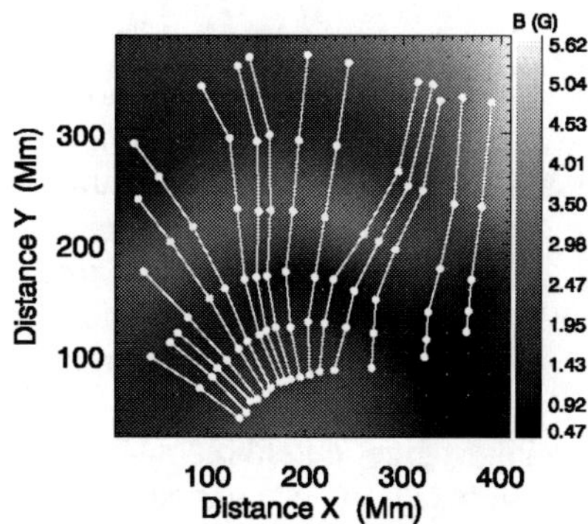

FIGURE 2. Magnetic map of the quite Sun obtained using an EIT wave observed by TRACE/EUV in 195 Å adapted from [26]

waves considered as fast MHD waves can also be used to determine the value of the radial component of the magnetic field at every location allowed by the observational precision. In this way, using the previously cited TRACE observations we can construct a magnetic map of the quiet Sun (see Figure 2), in other words EIT waves can serve as probes in a *magnetic tomography* of the quiet Sun. If points are joined across the lines, we will obtain the location of the EIT wavefront. Magnetic field varies between 0.47 and 5.62 G, however, these particular values should be handled with care as the interpolation will introduce spurious values at the two ends of the interval. It should be noted that this result has been obtained supposing a single value for density, in reality both magnetic field and density can vary along the propagation direction, as well. The method employed by [26] to find this magnetic map (magnetic field derived via the Alfvén speed) means that density and magnetic field cannot be determined at the same time. Further EUV density sensitive diagnostics line ratio measurements are required to establish a density map of the quiet Sun which will provide an accurate determination of the local magnetic field.

In conclusion, EIT waves propagating in the solar corona exhibit a wide range of applicabilities for plasma and field diagnostics. The fact that during their propagation EIT waves cover a large area of the solar surface (in the coronal) allows us to sample the magnetic field in the quiet Sun. EIT waves could serve as a link between eruptive events and localised oscillations, e.g. loop oscillations could be studied in a much broader context. Using a simple model we found that the minimum energy an EIT wave should have to produce a detectable loop oscillation is in the range of $10^{16} - 10^{19}$ J. This energy varies inversely proportional with the density of coronal loops.

Problems to be tackled in the future should include the study of attenuation of EIT waves with the aim of providing information about the magnitude of transport coefficients in the quiet Sun. The old problem of connecting different global waves still remain to be addressed.

Despite of the lack of high precision observations, EIT waves show a great potential for magneto-seismology of the solar corona.

CONCLUSIONS

The magnetic field in the solar atmosphere is one of the most important quantities which control the dynamics and stability of the plasma. Magnetic field also connects different layers of the solar atmosphere. The most fundamental result obtained from observations of the last few decades is that practically every solar feature that can be seen in the photosphere, chromosphere, corona, and beyond in the interplanetary space owes its existence, directly or indirectly, to the presence of the magnetic field (maybe the only exception is the photospheric granulation which has a purely hydrodynamic origin).

That is why the knowledge of the magnitude and structure of the magnetic field in the entire solar atmosphere is crucial. Magnetic field reconstructions based on extrapolation of photospheric field together with radio measurements have already proven to mimic the true nature of the field.

Significant advances in measuring the magnetic field and modelling its evolution and structure have been achieved using the indirect method of coronal seismology. Future high spatial and temporal resolution observations will further help the development of this fundamental diagnostic tool.

ACKNOWLEDGMENTS

The author would like to acknowledge the financial support received from the organizers of the meeting. The attendance to the meeting and the research carried out would not have been possible without the great support offered by the Nuffield Foundation (NUF-NAL 04) and NSF Hungary (OTKA, TO43741).

REFERENCES

1. M. Schüssler, in *The cosmic dynamo*, F. Krause, K.-H. Rädler, and G. Rüdiger (eds), Kluwer, Dordrecht, *IAU Symp.*, **138**, 27–32 (1993)
2. A.G. Kosovichev, J. Schou, P.H. Sherrer, R.S. Bogart, R.I. Bush et al., *Sol. Phys*, **170**, 43–61 (1997)
3. P. Caligary, F. Moreno-Insertis, and M. Schüssler, *Astrophys. J.*, **441**, 886–902 (1995)
4. A.R. Choudhuri and P. Gilman, *Astrophys. J.*, **316**, 788–800 (1987)
5. A.M. Title, and C.J. Schrijver, *The Xth Cambridge workshop on Cool Stars, Stellar Systems and the Sun*, ed. R.A. Donahue and J.A. Bookbinder, ASP Conf Ser., **154**, 345–351
6. R. Erdélyi, A. Kerekes, and N. Mole, *Astron. Astrophys*, **431**, 1083–1088
7. B. De Pontieu, R. Erdélyi, and S.P. James, *Nature*, **430**, 536–539 (2004)
8. R.G. Giovanelli, *Sol. Phys.*, **68**, 49–69 (1980)
9. S.K. Solanki and O. Steiner, *Astron. Astrophys.*, **234**, 519–529 (1990)
10. B. De Pontieu, T. Tarbell, and R. Erdélyi, *Astrophys. J.*, **590**, 502–518 (2003)
11. E.N. Parker, *Astrophys. J.*, **407**, 342–346 (1993)
12. I. Roussev, K. Dalsgaard, R. Erdélyi, and J.G. Doyle, *Astron. Astrophys.*, **370**, 298, (2001a)

13. I. Roussev, K. Dalsgaard, R. Erdélyi, and J.G. Doyle, *Astron. Astrophys.*, **375**, 228 (2001b)
14. I. Roussev, J.G. Doyle, K. Dalsgaard, and R. Erdélyi, *Astron. Astrophys.*, **380**, 719 (2001c)
15. B.J. Thompson, J.B. Gurman, W.M. Neupert, J.S. Newmark, J.-P. Delaboudinière, et al. 1999, *Astrophys. J.*, **517**, L151–L155 (1999)
16. Aschwanden, M.: *Physics of the Solar Corona*, Springer (2004)
17. H. Lin, J.R. Kuhn and R. Coulter, *Astrophys. J.*, **613**, 177–180 (2004)
18. P. Charvin, *Ann. d'Astrophys.*, **28**, 877–882 (1965)
19. V. Bommier, E.L. Degl'Innocenti, J.-L. Leroy and S. Sahal-Bréchot, *Sol. Phys*, **154**, 231–260 (1994)
20. , J. Lee, S.M. White, M.R. Kundu, Z. Mikic, and A.N. McClymont, *Astrophys. J.*, **510**, 413 (1999)
21. S.M. White, M.R. Kundu, V.I. Garaimov, T. Yokoyama, and J. Sato, *Astrophys. J.*, **576**, 505 (2002)
22. M.K. Bird and P. Edelhofer, in *Physics of the inner heliosphere*, R. Schwenn and E. Marsch (eds), Sringer, Berlin, Heidelberg, New York, 13–39 (1990)
23. Y. Uchida, *Pub. Astron. Soc. Jap.*, **22**, 341–364 (1970)
24. B. Roberts, P.M. Edwin, and A.O. Benz *Astrophys. J.*, **279**, 857–865 (1984)
25. V.M. Nakariakov, L. Ofman, E.D. DeLuca, B. Roberts, B. and J.M. Davila, *Science*, **285**, 862–864 (1999)
26. I. Ballai, *Sol. Phys*, in press (2007)
27. D. Banerjee, R. Erdélyi, R. Oliver, and E. O'Shea, E., *Sol. Phys*,in press, (2007)
28. R. Erdélyi, J.L. Ballester, and M.S. Ruderman, *Sol. Phys*, in press (2007)
29. C.E. DeForest, and J.B. Gurman, J.B. 1998, *Astrophys. J*, **501**, L217–L220 (1998)
30. D. Berghmans, and F. Clette *Sol. Phys.*, **186**, 207–229 (1999)
31. M.J. Aschwanden, L. Fletcher, C.J. Schrijver, and D. Alexander, *Astrophys. J.*, **520**, 880–894 (1999)
32. Y.M. Wang, 2000, *Astrophys. J.*, **543**, L89–L93 (2000)
33. B. De Pontieu, R. Erdélyi, and I. De Moortel, *Astrophys. J.*, **624**, 61–64 (2005)
34. I. Ballai, R. Erdélyi, and M. Goossens, *J. Plasma Phys.*, **64**, 235–247 (2000)
35. I. Ballai, and R. Erdélyi, *J. Plasma Phys.*, **67**, 79–97 (2002)
36. M.S. Ruderman and B. Roberts,: 2002, *Astrophys. J.*, **577**, 475– (2002)
37. M. Aschwanden, R.W. Nightingale, J. Andries, M. Goossens, M. and T. Van Doorsselaere, *Astropys. J.*, **598**, 1375–1386 (2003)
38. P.M. Edwin and B. Roberts, *Sol. Phys.*, **88**, 179–191 (1983)
39. Y. Taroyan, R. Erdélyi, T.J. Wang, and S.J. Bradshaw, *Astrophys. J.*, **659**, 173–176
40. M.J. Wills-Davey, and B.J. Thompson, 1999, *Sol. Phys.*, **190**, 467–483 (1999)
41. I. Ballai, R. Erdélyi, and B. Pintér *Astrophys. J. Lett*, **633**, L145–L148 (2005)
42. W. Chaplin, A.H. Dumbhill, Y. Elsworth, G.R. Isaak, C.P. McLeod, *et al.*, *Mon. Not. Roy. Astron. Soc.* **343**, 813–818 (2003)

Transverse Oscillations in Coronal Loops

I. Arregui, M. Luna, R. Oliver, J. Terradas, J. L. Ballester

Departament de Física. Universitat Illes Balears. E-07122. Palma de Mallorca (Spain)

Abstract. During last years, direct evidence about oscillations in different coronal structures has been obtained thanks to the detailed observations made by SoHO and TRACE. With the help of magnetohydrodynamic (MHD) wave theory, we can explain these observations as due to the excitation and propagation of MHD waves in the solar corona. In spite that there are many solar coronal structures in which oscillations have been detected (prominences, loops, plumes, coronal holes, etc.), in the following we will concentrate on transverse oscillations of coronal loops, reviewing some theoretical models developed to understand these oscillations in terms of MHD waves.

Keywords: Solar physics, MHD waves, Coronal Loops
PACS: 96.60.-j, 96.50.Tf, 96.60.pf

INTRODUCTION

From the 1960s we had already indirect evidences of coronal oscillations coming from radio observations and coronal emission lines. SoHO and TRACE satellites have provided us with a lot of evidences about the presence of oscillations in coronal loops, in particular about transversal oscillations in coronal loops ([1], [2], [3]) and their short time damping; about compressive waves in short and long coronal loops ([4], [5], [6]); and about quasi-periodic oscillations of the Doppler shift and intensity of the iron XIX and XXI emission lines in coronal loops ([7], [8]) and their damping.

Theoretical studies of oscillations and MHD waves in the solar corona started more than twenty years ago, before many of the above mentioned observations, and the interest of their study lies in their potential relationship with the coronal heating problem. ([9], [10]) gave a tentative explanation of the indirect evidences of coronal oscillations in terms of standing or impulsive MHD waves in coronal loops considered as straight and infinite magnetic flux tubes. Nowadays, the theoretical range of periods is well covered by the time resolution available in ground-based telescopes and in onboard satellite instruments and this fact, together with other observational details, allows the attribution of theoretical counterparts (modes of oscillation) to observations. Then, oscillations observed in coronal loops have been interpreted in terms of MHD waves, and different MHD modes responsible for those oscillations have been identified ([11], [12], [13], [14]). The identification of these modes is the key for the local seismology of the solar corona and it has allowed preliminary estimations of the magnetic field and plasma β in coronal loops [13]. The main goal of the local seismology of the solar corona is to use the oscillations detected in coronal structures, and their interpretation in terms of MHD waves, as a tool to perform a diagnostic of the physical properties of those structures and in the near solar corona. For instance, the coronal magnetic field in the magnetic structures and surroundings, the transport coefficients (viscosity, resistivity,

CP934, *Flows, Boundaries, Interactions*
edited by C. Dumitrache, V. Mioc, and N. A. Popescu

thermal conductivity), etc.

DAMPED FAST KINK-MODE OSCILLATIONS IN A STRAIGHT LOOP

In 1998, the imaging telescope onboard TRACE detected damped transversal oscillations of coronal loops located within an active region ([1], [2], [15]). Following this observation, other observations of transversal oscillations in coronal loops have been obtained with TRACE ([16], [17], [18]). In most cases, oscillations are triggered by a nearby flare [19] and produce transversal loop motions. These oscillations have been interpreted as standing fast kink magnetohydrodynamic modes of the loops since they produce transversal displacements of the axis. The average period of the oscillating loops is 321 ± 140 s while the average damping time is 580 ± 385 s.

The interpretation of the physical mechanism that produces these oscillations is of great interest since it can provide with indirect information about the coronal plasma and in addition it can shed new light on the role of wave heating in coronal loops. Several mechanisms of wave damping have been proposed, the most popular ones being phase mixing and resonant absorption. Ofman and Aschwanden [20] used observational data to calculate the scaling laws of different damping mechanisms and suggested that phase mixing can be favored in front of resonant absorption. However, phase mixing in the simple form derived by [21] and used to calculate the damping times indicates that the magnetic Reynolds number (R_m) should be five or six orders of magnitude smaller than the classical coronal value ($R_m \approx 10^{12}$) to reproduce the damping times of the oscillating loops. It has been suggested that such a small anomalous Reynolds numbers could be produced by turbulence. On the other hand, Goossens et al. [22] pointed out that the damping of kink oscillations by resonant absorption gives a consistent explanation of the rapid decay of the observed coronal loop oscillations if the transverse inhomogeneity length-scale is allowed to vary from loop to loop. This mechanism was used by [23] to show that the damping times for the kink mode oscillations are in agreement with the observations and that resonant absorption can provide with a new diagnostic tool for the density contrast in loops.

Terradas et al. [25] studied the time-dependent problem for thin and also thick boundaries. The equilibrium configuration used to model a coronal loop of radius R and length L consists of a homogeneous straight tube described in cylindrical coordinates, with ρ_i the density inside the loop, ρ_e the gas density in the loop environment ($\rho_i > \rho_e$), and l the width of the inhomogeneous layer. The loop is permeated by a vertical and uniform magnetic field ($\mathbf{B} = B_0\mathbf{e}_z$). The Alfvén speed, $v_A = B_0/\sqrt{\mu_0\rho_0}$, takes the value v_{Ai} inside the loop and increases smoothly to the value v_{Ae} in the surrounding coronal medium (see Figure 1). Linear perturbations about this equilibrium in the zero-β limit can be readily described using the magnetohydrodynamic equations with constant resistivity and they concentrate on the fundamental mode, with $k_z = \pi/L$, so the photospheric line-tying effect is automatically incorporated in the model. They also impose $m = 1$, i.e. they excite the kink mode, which is the only oscillation that displaces the tube axis.

Coronal oscillations are often produced by an impulsive event and a time-dependent simulation is more appropriate to describe the evolution of the system. To study the effect

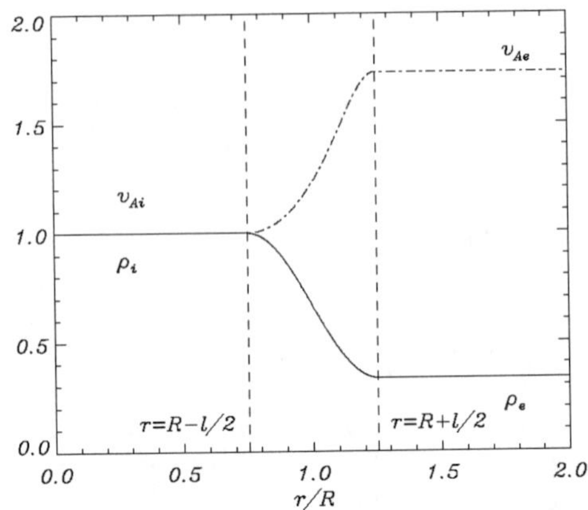

FIGURE 1. Dependence of the density (continuous line) and the Alfvén speed (dot-dashed line) on the radial coordinate, r. The radius of the loop is R and the width of the inhomogeneous layer is l. The vertical dashed lines mark the boundaries of the inhomogeneous layer. The density and the Alfvén speed are normalized to the value at the loop axis ($r = 0$). Distances are normalized to R. In this figure $\rho_i/\rho_e = 3$ and $l/R = 0.5$.

of an arbitrary initial perturbation they have excited a typical loop with a disturbance located outside it. The initial perturbation induces disturbances that propagate towards the loop and also wavefronts traveling in the opposite direction. Once the perturbations reach the loop position some of the energy of the wave packet is deposited in the tube. The amount of energy deposited in the loop depends on the shape and distance of the initial perturbation from the tube. Since we are primarily interested on how the energy trapped by the loop is transferred to the inhomogeneous layer by resonant absorption, we have used an initial disturbance that deposits a considerable amount of energy in the loop. Figure 2 shows the radial velocity component, v_r, at the loop center ($r = 0$) as a function of time. We can clearly identify several phases. There are initially two large extrema followed by short-period oscillations which are quickly attenuated (see inner plot of Figure 2). After these oscillations ($t > 10\tau_A$) there is a long-period oscillation with an amplitude that attenuates in time. The short-period oscillatory phase is related with the excitation of a leaky mode, which is a radially propagating fast wave. After the short leaky transient the loop oscillates with a much longer period. The amplitude of this mode is attenuated due to the conversion of energy from the global kink mode to the Alfvén modes. We have fitted an exponential function of the form $\exp(-t/\tau_d)$ to the envelope of the signal and have numerically determined the damping time, τ_d. It is important to point out that the resonant absorption mechanism acts very quickly (see Figure 2).

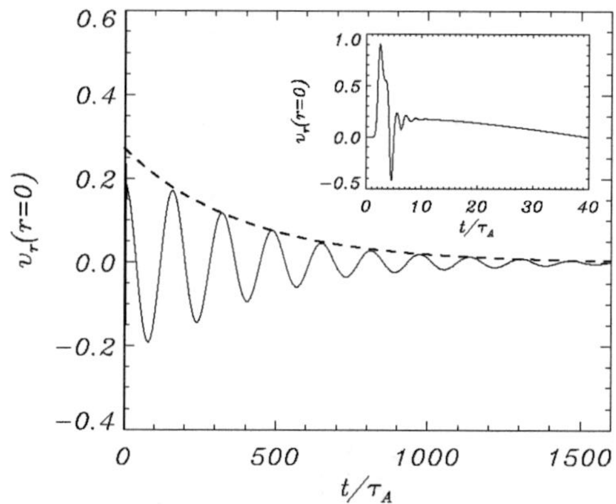

FIGURE 2. Plot of v_r at the center of the loop as a function of time. After a short leaky transient (for $t/\tau_A \leq 10$), detailed in the inner plot, the loop oscillates with the corresponding quasi-mode. The dashed line is a fit of the form $\exp(-t/\tau_d)$.

DAMPED FAST KINK-MODE OSCILLATIONS IN A CURVED LOOP

The equilibrium configuration used to model a coronal loop of radius R and length L consists of a curved semicircular tube (Figure 3). The loop has a circular cross-section, its center is located at $r = r_0$ and its length is $L = \pi r_0$, where r is the distance from the axis of the torus (rather than the distance from the loop center as in the straight cylinder). The loop and the coronal environment are permeated by a purely toroidal magnetic field of the form $\mathbf{B} = B_0 r_0 / r\, \mathbf{e}_\phi$. It has been assumed that the background density profile is of the form $\rho = \rho_0 (r/r_0)^\alpha$, α being an arbitrary parameter that allows to study different density profiles with height. The coronal loop is modelled as a density enhancement of the background profile and in the xz-plane is as follows (the density is independent of ϕ and so has the same form in other planes containing the x-axis),

$$\rho = \begin{cases} \rho_i \left(\frac{r}{r_0}\right)^\alpha, & 0 \leq \sqrt{x^2 + (r - r_0)^2} \leq R, \\ \rho_e \left(\frac{r}{r_0}\right)^\alpha, & \sqrt{x^2 + (r - r_0)^2} > R. \end{cases}$$

The results presented in [26] correspond to a density contrast between the loop and the corona $\rho_i/\rho_e = 3$. To study the damping by resonant absorption the density changes smoothly from the loop to the corona over a characteristic distance l. Linear perturbations about this equilibrium in the zero-β limit are described using the MHD equations in toroidal coordinates, and the two-dimensional eigenvalue problem has been solved numerically.

One of the differences of the oscillation modes in the curved loop with respect to the straight cylindrical tube is that curvature introduces preferential directions of oscillation. The polarisation of motions is similar to that of the kink mode, basically producing a lateral displacement (now horizontal or vertical with respect to the photosphere) of the whole tube. A significant feature of the two kink modes in a torus is that their frequencies are very similar.

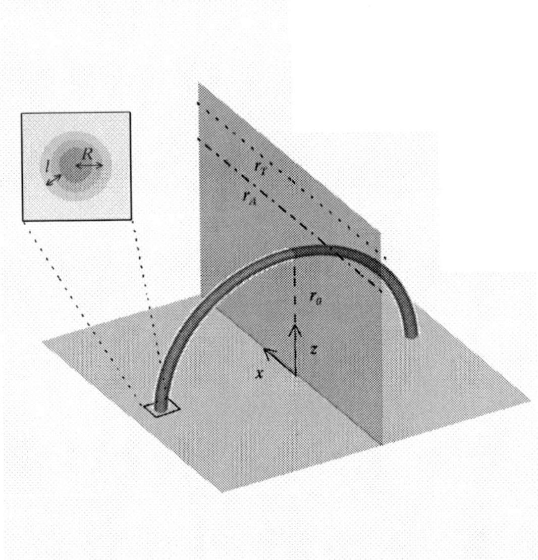

FIGURE 3. Sketch of the equilibrium configuration, made of a loop whose magnetic field is in the toroidal direction. The major radius of the tube is r_0, so its length is $L = \pi r_0$. The minor radius of the loop is R (see detail of the loop cross-section) and has an inhomogeneous layer of width l. The photosphere is located at $z = 0$ and the x-axis coincides with the axis of the torus. The positions $r = r_A$ and $r = r_T$ mark the external resonant position (dot-dashed line) and the end of the tunneling barrier (dotted line) for an equilibrium with $\alpha > -4$.

Next, the variation of the frequency ($\omega = \omega_R + i\omega_I$) with the thickness of the resonant layer for a loop with $R/L = (20\pi)^{-1} \simeq 0.016$ has been studied. The results show that ω_R deviates very little (less than 0.2%) from the thin tube value in the considered range ($0.1 \leq l/R \leq 0.5$), while ω_I differs from the thin tube thin boundary (TTTB) approximation [24] by 12% at most. This discrepancy can be well appreciated in Figure 4 by comparing the triangles (corresponding to the horizontal kink mode of the torus) with the solid line (TTTB value). This plot reveals that there is only a small difference in the imaginary part of the frequency compared to the TTTB value and that, because of the inherent limitations of the TTTB formula, this difference grows as the thickness of the transition layer is increased. Hence, even in this curved cylinder (with $\alpha = -4$) ω_I is essentially proportional to l/R.

Now let us discuss more precisely the influence of the density profile (through the power index α) on the frequency of the two kink modes. Since the variations in ω_R are so small, we concentrate on the imaginary part of the frequency, which is plotted in Figure 4 versus l/R for $\alpha = -5, -4, -3$. Solutions for $\alpha \neq -4$ display two distinct behaviors: for $l/R \geq 0.2$, ω_I varies almost linearly with the dimensionless thickness of the resonant layer, closely following the rough TTTB expression; for $l/R \leq 0.2$, however, the results slightly depart from this approximately linear trend in such a way that the damping time decreases a little. These results must be interpreted as follows: resonant absorption in the layer between the loop and the coronal environment governs the damping time of the two kink modes for $l/R \geq 0.2$ and only for very thin transition layers do the external resonant coupling and the wave tunneling have some effect on the mode damping.

The first main result is that curvature and density structuring introduce preferential directions of oscillation showing the existence of a vertically and a horizontally polarized kink mode which are very difficult to distinguish from the observational point of view. Also, it is shown that the period and damping time of a curved toroidal loop surrounded by a smooth transition layer are similar to those of a straight cylinder with the same density transition to the corona especially in the limit of thin tube and thin boundary. Thus, the oscillatory and damping features of the kink modes are determined basically by the loop length, density contrast and width of the inhomogeneous layer. This is still true for values of α different from -4.

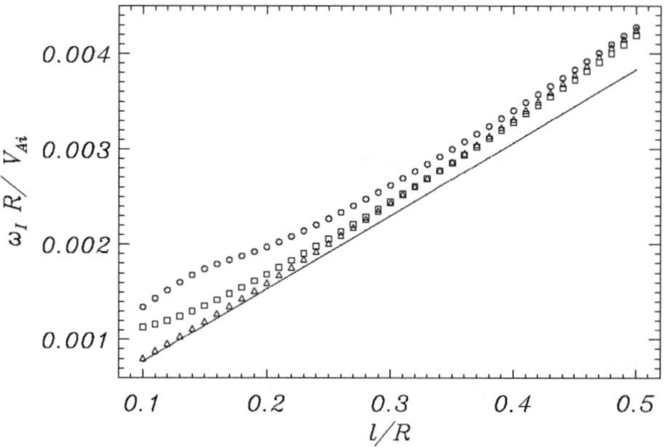

FIGURE 4. Imaginary part of the frequency as a function of the width of the inhomogeneous layer for the kink mode with horizontal velocity polarization. The solid line corresponds to the thin loop, thin boundary approximation, whereas circles, triangles and squares correspond to $\alpha = -3, -4, -5$. In this plot $R/L = (20\pi)^{-1} \simeq 0.016$.

COLLECTIVE OSCILLATIONS IN CORONAL LOOPS

Observational evidence about transversal coronal loop oscillations in coronal arcades has been also obtained. [3] studied the features of the oscillating loops forming an arcade, and the observations suggest that the loops do not oscillate independently and that different loops oscillate following an organized motion. [27] have analyzed the normal modes and the time evolution of the collective oscillations of two coronal loops modeled as smoothed, dense plasma slabs in a uniform magnetic field (Figure 5).

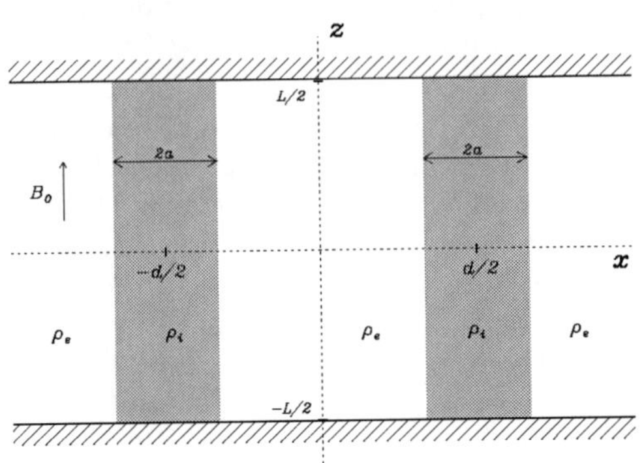

FIGURE 5. Sketch of the two-slab system. The shaded area represents the density enhancement of the two slabs while the hatched area represents the photospheric medium, that fixes the feet of the slabs and produces the line-tying effect.

The normal mode analysis indicates that the symmetric mode is the only trapped mode for any distance between slabs, while the antisymmetric mode is leaky for small slab separations. However, there is a large range of slab separations (larger than the critical distance, $d \approx 1.04L$) where both trapped modes coexist and possess very close frequencies. Next, the temporal evolution of symmetric, antisymmetric and arbitrary excitations, for a typical loop system, have been studied. They have found that for symmetric disturbances, and after a brief transient, all that remains is the undamped trapped mode, with energy confined to both slabs. On the other hand, since there are no trapped antisymmetric modes for slab separations smaller than the critical one, an antisymmetric-like initial disturbance can only deposit energy in the leaky antisymmetric modes. The excitation of the fundamental antisymmetric trapped mode is only possible for distances greater than the critical one.

An arbitrary excitation, when the distance between the slabs is greater than the critical distance, leads to the simultaneous excitation of the symmetric and antisymmetric modes. Since their frequencies are quite similar, the oscillations do not attain a constant amplitude and show a sinusoidal modulation. This is a well known collective beating phenomenon which is completely equivalent to the behaviour of two weakly coupled oscillators.

A significant improvement of this model is to consider two cylindrical loops instead of Cartesian slabs. Then, [28] have studied fast magnetohydrodynamic waves in a system of two coronal loops modeled as smoothed, dense plasma cylindrical flux tubes in a uniform magnetic field (Figure 6). The collective properties of the system due to the interaction between the individual loops have been analyzed from two points of view. Firstly, the normal modes of the equilibrium configuration have been obtained and the dependence of the frequency with the separation between tubes as well as the spatial distribution of the eigenfunctions have been investigated. Secondly, they have analyzed the time dependent problem of the excitation of a pair of tubes. Four collective fundamental trapped modes that produce transverse oscillations of the loops have been found. The velocity field is more or less uniform in the interior of the loops, and the loops move basically as a solid body. The four velocity field solutions have definite symmetry with respect to the y-axis. In one case, the velocity field within the tubes lies in the x-direction and it is symmetric with respect to the y-axis. We call this mode S_x where S refers to the symmetry of the velocity field, i.e. they move in phase, and the subscript x refers to the direction of the velocity within the tube. In the second case, the velocity is antisymmetric with respect to the y-axis. We call this mode A_x. Similarly, when the velocity lies in the y-direction and it is symmetric or antisymmetric with respect to the y-axis, these modes have been called these modes S_y and A_y, respectively. The excitation of these modes depends on the shape and location of the initial disturbance. We have also found that the loop pair oscillates with the normal modes after an initial disturbance. In some cases, the system shows beating and the phase lag between the loops is $\pi/2$.

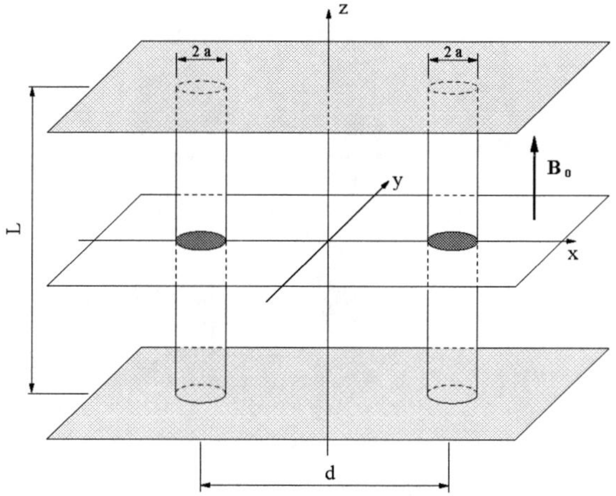

FIGURE 6. Model of two straight cylinders of radius a and internal density ρ_i, embedded in the coronal medium of density ρ_e, where $\rho_e < \rho_i$. The distance between the loop centers is d and the tube length is L. The magnetic field is uniform in both media and points along the $z-$axis. The loop feet are anchored in the photosphere, modelled as two surfaces in this geometry (plotted as two shaded areas). The xy-plane is represented as a white slice.

THE INFLUENCE OF INTERNAL STRUCTURE OF CORONAL LOOPS ON THEIR TRANSVERSE OSCILLATIONS

High-resolution observations obtained by TRACE indicate the existence of sub-resolution transverse structuring which has not been taken into account in theoretical models developed to study transversal coronal loop oscillations. Then, [29] have tried to assess the effect of the possibly unresolved internal structure of a coronal loop on the properties of its transverse oscillations and on the efficiency of resonant absorption as a damping mechanism of these oscillations. The equilibrium magnetic and plasma configuration of a single coronal loop has been modeled by means of a one-dimensional system of two identical parallel over-dense slabs in Cartesian geometry. The magnetic field is straight and pointing in the z-direction, $\vec{B} = B\hat{e}_z$ (Figure 7). Then, the period of the oscillation and the damping time have been computed for the resonantly damped fundamental kink mode. These quantities have then been compared to those obtained for two models for a single equivalent slab without internal density structuring. It has been found that the period and the damping time of a coronal loop with internal density structuring change by less than 15%, when compared to the same oscillatory properties of a single coronal loop with either the same density contrast or a single coronal loop with the same total mass. Therefore the internal density structuring of a coronal loop does not affect its oscillatory properties very much. However, the sub-resolution structuring of a coronal loop with different densities in its components or with different widths could vary these results.

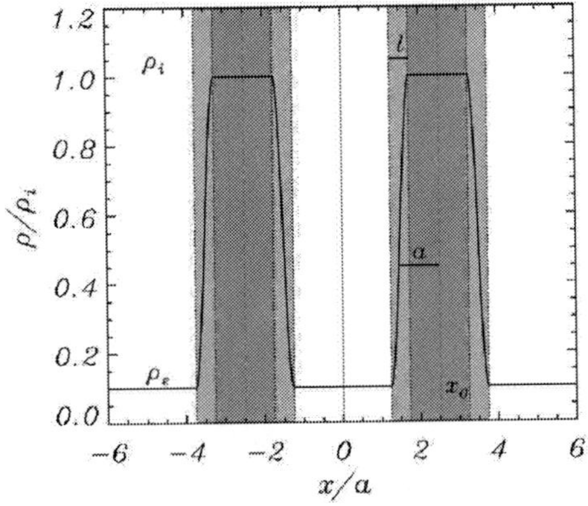

FIGURE 7. Schematic representation of the two density enhancements (shaded regions) of half-width a, centered on $\pm x_0$ and representing a coronal loop composed of two slabs in the direction transverse to the equilibrium magnetic field. These enhancements with internal density ρ_i connect to the external medium, with density ρ_e, by transitional non-uniform layers (light-shaded regions) of thickness l.

CONCLUDING REMARKS

An extended observational background about oscillations in quiescent solar prominences, coronal loops, plumes, coronal holes, etc. has been gathered during last years. However, to make further progress new two-dimensional high-resolution observations, powerful data analysis techniques and theoretical developments are needed. The development of more sophisticated numerical simulations to study the time evolution of the oscillations of more realistic equilibrium configurations (3D, curvature, inhomogeneities, etc.), including also further physical effects (resonant absorption, non adiabatically, phase mixing, etc.), is desirable. Once realistic configurations have been considered, the obtained results will allow: (a) A better comparison with the observations; (b) A further refinement of the theoretical models; (c) To obtain reliable estimations of solar coronal parameters; and (d) To obtain information about the contribution to the coronal heating of the different mechanisms responsible of the damping of observed oscillations.

ACKNOWLEDGMENTS

The authors acknowledge financial support from MEC under grant AYA2006-07637.

REFERENCES

1. M. J. Aschwanden, L. Fletcher, C. Schrijver and D. Alexander, *The Astrophysical Journal*, **520**, 880–894 (1999).
2. V. M. Nakariakov, L. Ofman, E. E. Deluca, B. Roberts and J. M. Davila, *Science*, **285**, 862–864 (1999).
3. E. Verwichte, V. M. Nakariakov, L. Ofman and E. E. Deluca, *Solar Physics*, **223**, 77–94 (2004).
4. D. Berghmans and F. Clette, *Solar Physics*, **186**, 207–229 (1999).
5. R. W. Nightingale, M. J. Aschwanden and N. E. Hurlburt, *Solar Physics*, **190**, 249–265 (1999).
6. I. De Moortel, J. Ireland and R. Walsh, *Astronomy and Astrophysics*, **355**, L23–L26 (2000).
7. B. Kliem, I. E. Dammasch, W. Curdt and K. Wilhelm, *The Astrophysical Journal*, **568**, L61–L65 (2002).
8. T. J. Wang, S. K. Solanki, W. Curdt, D. E. Innes and I. E. Dammasch, *The Astrophysical Journal*, **574**, L101–L104 (2002).
9. B. Roberts, P. Edwin and A. O. Benz, *Nature*, **305**, 688–690 (1983).
10. B. Roberts, P. Edwin and A. O. Benz, *The Astrophysical Journal*, **279**, 857–865 (1984).
11. B. Roberts, *Solar Physics*, **193**, 139–152 (2000).
12. B. Roberts, in *Proc. of SOHO 13 "Waves Oscillations and Small-Scale Transient Events in the Solar Atmosphere: A Joint View from SOHO and TRACE"*, eds. R. Erdélyi, J. L. Ballester and B. Fleck. ESA SP-547, 2004, pp. 1–14.
13. M. J. Aschwanden, *Physics of the Solar Corona*, Springer-Praxis (2005).
14. V. M. Nakariakov and E. Verwichte, *Living Reviews in Solar Physics*, **2**, 3 (2005).
15. C. J. Schrijver, A. M. Title, T. E. Berger, L. Fletcher, N. E. Hurlburt, R. W. Nightingale, R. A. Shine, T. D. Tarbell, J. Wolfson, L. Golub, J. A. Bookbinder, E. E. DeLuca, R. A. McMullen, H. P. Warren, C. C. Kankelborg, B. N. Handi and B. De Pontieu, *Solar Physics*, **187**, 261–302 (1999).
16. C. J. Schrijver, M. J. Aschwanden and A. M. Title, *Solar Physics*, **206**, 69–98 (2002).
17. M. J. Aschwanden, B. De Pontieu, C. J. Schrijver and A. M. Title, *Solar Physics*, **206**, 99–132 (2002).
18. V. M. Nakariakov and L. Ofman, *Astronomy and Astrophysics*. **372**, L53–L56 (2001).
19. H. S. Hudson and A. Warmuth, *The Astrophysical Journal*, **614**, 85–88 (2004).
20. L. Ofman and M. J. Aschwanden, *The Astrophysical Journal*, **576**, L153–L156 (2002).
21. J. Heyvaerts and E. R. Priest, *Astronomy and Astrophysics*, **117**, 220–234 (1983).
22. M. Goossens, J. Andries and M. J. Aschwanden, *Astronomy and Astrophysics*, **394**, L39–L42 (2002).
23. M. J. Aschwanden, R. W. Nightingale, J. Andries, M. Goossens and T. Van Doorsselaere, *The Astrophysical Journal*, **598**, 1375–1386 (2003).
24. T. Van Doorsselaere, J. Andries, S. Poedts and M. Goossens, *The Astrophysical Journal*, **606**, 1223–1232 (2004).
25. J. Terradas, R. Oliver and J. L. Ballester, *The Astrophysical Journal*, **642**, 533–540 (2006).
26. J. Terradas, R. Oliver and J. L. Ballester, *The Astrophysical Journal*, **650**, L91–L94 (2006).
27. M. Luna, J. Terradas, R. Oliver and J. L. Ballester, *Astronomy and Astrophysics*, **457**, 1071–1079 (2006).
28. M. Luna, J. Terradas, R. Oliver and J. L. Ballester, *In preparation*, (2007).
29. I. Arregui, J. Terradas, R. Oliver and J. L. Ballester, *Astronomy and Astrophysics*, **466**, 1145–1151 (2007).

Thermally Damped Waves in the Solar Corona

A. Marcu* and I. Ballai†

*Department of Theoretical and Computational Physics, Babes-Bolyai University, M.Kogalniceanu,1, Cluj-Napoca, Romania
†Solar Physics and Space Plasma Research Centre , Dept. of Applied Mathematics, University of Sheffield,Hicks Building, Hounsfield Road, Sheffield, S3 7RH, England (UK)

Abstract.
Waves observed by high resolution telescopes in the solar corona are subject to a rapid damping which could be due to non-ideal effects and/or transversal inhomogeneities. The attenuation of modes is currently used to determine, e.g. the value and structure of coronal magnetic field, a fundamental parameter which cannot be measured directly. Here we study the damping of linear MHD modes propagating in a stratified plasma in the presence of thermal conduction. For the chosen particular equilibrium we show the importance of the transversal motion in the process of wave damping.

Keywords: MHD, waves
PACS: 96.60.P

INTRODUCTION

The launch of the high resolution satellites and the study of data provided by these space telescopes, revealed to solar scientists the existence of waves propagating in the solar corona in the magnetohydrodynamic (MHD) frequency domain. Waves and oscillations have been observed in various coronal structures such as polar plumes (Ofman et al. 1997, DeForest and Gurman 1998) and coronal loops (Berghmans and Clette 1999, De Moortel et al. 2000) and in the quiet Sun (Thompson et al. 1999, Ballai et al. 2005). Special attention has been paid to ducted waves in coronal loops because they carry information about field structure and plasma transport parameters. Using the technique of coronal seismology magnetic field values have been obtained in coronal loops and quiet Sun (e.g. Nakariakov et al. 1999, Ballai 2007).

The common feature of the observed waves and oscillations in the solar corona is their rapid damping. De Moortel and Hood (2003 and 2004) studied the damping of coronal compressional waves supposing an 1D model. Comparing the observational damping values with theoretical results (in this model the damping is due to non-ideal effects, stratification and field line opening) they found that the propagating slow MHD modes can be damped and there is a minimum damping time (or length) that can be obtained by thermal conduction alone. They concluded that if a strong damping is required a new ingredient must to be included. Combining the effect of gravitational stratification, thermal conduction and optical thick radiation, De Moortel and Hood (2004) showed that the stratification increase the damping length considerably. Their numerical analysis proved that the combined action of thermal conduction and loop cross-area divergence yields damping length values in a very good agreement with TRACE observations.

CP934, *Flows, Boundaries, Interactions*
edited by C. Dumitrache, V. Mioc, and N. A. Popescu

In this paper we have investigated, in 2D model, the effect of gravitational stratification on thermally damped linear compressional waves propagating in coronal loops.

THE MODEL AND BASIC EQUATIONS

We suppose a relatively simplified model in which the waves propagate upwards in low-beta completely ionized gas subjected to a constant gravitational acceleration.The observed wavelengths and periods are large compared to the ion Larmor radius and the period of gyration around a magnetic field line, so these waves can be studied within the MHD framework.

In order to simplify the model we restrict our model only to those waves which have wavelengths comparable to the size of the loop (waveguide). In this way we neglect any dispersive effect which would appear due to the geometrical transversal size of the coronal loop. It was shown by Edwin and Roberts (1983) that waves corresponding to this limit (know as the wide tube limit) are weakly dispersive, i.e. they propagate similar to waves in an unbounded plasma. We suppose that the vertical wavelength of the waves is comparable to the gravitational wavelength, so the effect of gravitational stratification must be taken into account. In order to describe non-ideal effects, which may act to damp the considered modes, we consider thermal conduction as one of the most plausible mechanisms that can affect the propagation of MHD waves (certainly for slow MHD modes). Under coronal conditions, thermal conductivity is a tensorial quantity with the parallel component to the ambient magnetic field being much larger than the perpendicular one (Ruderman et al. 2000).

The dynamics of the plasma is described by the system of MHD equations

$$\frac{D\rho}{Dt}+\rho\nabla\cdot\mathbf{v}=0, \tag{1}$$

$$\rho\frac{D\mathbf{v}}{Dt}=-\nabla p+\frac{1}{\mu}\left[(\nabla\times\mathbf{B})\times\mathbf{B}\right]+\rho\mathbf{g}, \tag{2}$$

$$\frac{\partial\mathbf{B}}{\partial t}=\nabla\times(\mathbf{v}\times\mathbf{B}), \tag{3}$$

$$\frac{\rho^{\gamma}}{\gamma-1}\frac{D}{Dt}\left(\frac{p}{\rho^{\gamma}}\right)=-\nabla\cdot\mathbf{q},\quad D/Dt=\partial/\partial t+\mathbf{v}\cdot\nabla. \tag{4}$$

Here, $\mathbf{v}$ and $\mathbf{B}$ are the velocity and magnetic induction vectors, p, ρ and $\mathbf{g}=-g\mathbf{z}$ are the pressure, density and the gravitational acceleration and γ denotes the adiabatic index. The constant equilibrium magnetic field is pointing in the positive z direction (upward). The perturbations of the magnetic field and velocity are denoted by $\mathbf{b}=(b_x,0,b_z)$ and $\mathbf{v}=(v_x,0,v_z)$. In Eq. (4) $\mathbf{q}$ signifies the heat flux due to thermal conduction. Taking into account only the parallel component of the thermal conductivity, the heat flux can be written as (see, e.g. Braginskii 1965): $\mathbf{q}=-\kappa_{\parallel}\mathbf{b}\left(\mathbf{b}\cdot\nabla T+\frac{b_z}{B_0}\frac{dT_0}{dz}\right)$, where $\kappa_{\parallel}$ is the parallel component of the thermal conductivity, $\mathbf{b}=\mathbf{B}_0/B_0$ is unit vector in the direction of the equilibrium magnetic field, T_0 and T are the equilibrium and perturbed temperatures related by $\frac{p}{p_0} = \frac{\rho}{\rho_0} + \frac{T}{T_0}$. The coefficient of thermal conductivity, $\kappa_\parallel$, is given by $\kappa_\parallel = \frac{3\rho_0 k_B^2 T_0 \tau_e}{m_p m_e}$, where k_B is the Boltzmann constant, τ_e is the electron collision time, m_p and m_e are the proton and electron masses and ρ_0 is the unperturbed density of the plasma. For typical coronal conditions, $\kappa_\parallel \approx 5\times 10^4$ m s^{-3} kg K^{-1} (Ruderman et al. 2000).We consider small but finite amplitude perturbations about the equilibrium, i.e. we write all physical quantities (except the velocity) in the form of $f = f_0 + \tilde{f}$, where f_0 is the equilibrium quantity and $\tilde{f}$ is its Eulerian perturbation.

Observations show that the amplitudes of the modes are very small (De Moortel et al. 2002), therefore we will limit our description to linear waves only, so we consider small but finite amplitude perturbations about the equilibrium, i.e. in the governing equations we neglect every product of perturbations.

The plasma is in hydrostatic equilibrium expressed by the condition

$$\frac{dp_0}{dz} = -\rho_0 g, \tag{5}$$

which means that the density and Alfvén wave profiles are given by

$$\rho_0(z),\ v_A(z) = \begin{cases} \rho_0(0)e^{-z/H}, \\ v_{A0}(0)e^{z/2H}, \end{cases} \tag{6}$$

where $\rho_0(0)$ and $v_{A0}(0)$ are the density and Alfvén speed at $z=0$ and $H(=c_S^2/\gamma g)$ is the isothermal scale-height.

We consider a y-independent harmonic solution of the form $f(x,z,t) = f(z)e^{i(\omega t - k_\perp x)}$, where ω is the frequency of the waves and $k_\perp$ is the component wave vector in the x direction. The system of equations (1)–(4) can be reduced to two coupled ordinary differential equations (ODE) for v_x and v_z

$$\left[\omega^2 - \frac{k_\perp^2 c_S^2}{\gamma} - H\left(\omega^2 - k_\perp^2 v_A^2(z)\right)\frac{d}{dz} - v_A^2(z)H\frac{d^3}{dz^3}\right] v_x =$$

$$ik_\perp H\left[\left(\omega^2 - \frac{c_S^2}{H^2\gamma}\right) + \frac{c_S^2}{\gamma H}\frac{d}{dz}\right] v_z, \tag{7}$$

and

$$\left[\omega^2 - k_\perp^2(c_S^2 + v_A^2(z)) + \frac{i\chi\gamma(\omega^2 - k_\perp^2 c_S^2 - k_\perp^2 v_A^2(z))}{\omega H(\gamma-1)}\frac{d}{dz} + \right.$$

$$v_A^2\frac{d^2}{dz^2} - \frac{i\chi\gamma\omega}{\gamma-1}\frac{d^2}{dz^2} + \frac{i\chi k_\perp^2(cs^2 + v_A^2(z))}{\omega(\gamma-1)}\frac{d^2}{dz^2} -$$

$$\left. -\frac{i\chi\gamma v_A^2(z)}{H\omega(\gamma-1)}\frac{d^3}{dz^3} - \frac{i\chi\gamma v_A^2(z)}{\omega(\gamma-1)}\frac{d^4}{dz^4}\right] v_x = c_S^2 \times$$

$$\left[-\frac{ik_\perp}{\gamma H} + \left(ik_\perp + \frac{\chi\gamma k_\perp}{\omega H^2(\gamma-1)}\right)\frac{d}{dz} - \frac{2\chi\gamma k_\perp}{\omega H(\gamma-1)}\frac{d^2}{dz^2} + \frac{\chi\gamma k_\perp}{\omega(\gamma-1)}\frac{d^3}{dz^3}\right] v_z. \tag{8}$$

where $\chi = \frac{m_p(\gamma-1)^2\kappa_\parallel}{2\gamma k_B \rho_0}$. Let us introduce the dimensionless quantities

$$a = k_\perp H, \quad \beta = \frac{\omega H}{c_S}, \quad \alpha = \frac{\chi}{H^2\omega}, \tag{9}$$

and the new variable

$$\eta = \frac{c_S^2\beta^2 \exp(-z/H)}{v_{A0}^2} = \frac{\omega^2 H^2 \exp(-z/H)}{v_{A0}^2}, \tag{10}$$

where, physically, the new variable η signifies the ratio between the Alfvén transit time for crossing the gravitational scale height and the wave perturbation period.

The z-dependence of the coefficient operators in the system of equations (7)-(8) does not allow us to determine v_z directly. Instead, using that the operators of v_z are z-independent (except derivatives), we can derive a solution for v_x and then the solution for v_z can be found based on the result obtained for v_x. Accordingly, the system of equations (7)-(8) can be reduced to

$$\sum_{n=0}^{6} \psi_n(\eta)\frac{d^n v_x}{d\eta^n} = 0, \tag{11}$$

where the coefficients ψ_i $(i = 1,..,6)$ can be easily calculate and were given in Ballai et al. (2007)

RESULTS AND DISCUSSION

In the particular case of waves in unstratified plasma under the effect of thermal conduction, the system of equations describing the evolution of waves is simplified by considering $H \to \infty$, i.e. the coefficient operators in the the system (7)-(8) reduce to forms which allow a direct investigation of waves. Since the effect of gravitational stratification is neglected, the Alfvén speed becomes constant. Due to the particular choice of the model, all perturbations can be Fourier-analyzed in the z direction and write all quantities proportional to $\exp(-ik_z z)$. The system of equations describing the dynamics of waves can easily derived to lead to the dispersion relation

$$\omega^5 + F_1\omega^4 + F_2\omega^3 + F_3\omega^2 + F_4\omega + F_5 = 0, \tag{12}$$

where the coefficients F_i, $(i = 1\ldots5)$ are defined as

$$F_1 = \frac{i\chi\gamma k_z^2}{\gamma-1}, \quad F_2 = -K^2(v_A^2 + c_S^2), \quad F_3 = -\frac{i\chi k_z^2 K^2(c_S^2+v_A^2)}{\gamma-1},$$

$$F_4 = k_z^2 K^2 c_S^2 v_A^2, \quad F_5 = \frac{i\gamma\chi c_S^2 v_A^2 K^2 k_z^4}{\gamma-1}.$$

where $K^2 = k_\perp^2 + k_z^2$.The solutions of (12) describe the frequency of slow, fast MHD and thermal modes. Under coronal conditions $c_S^2 \ll v_A^2$ and the slow waves represent

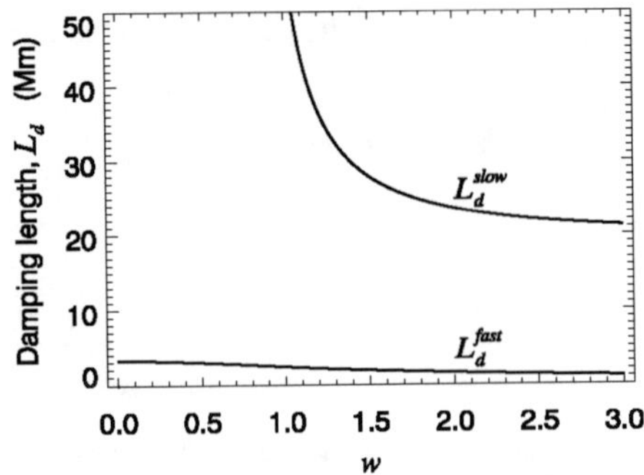

FIGURE 1. The variation of the damping length of slow and fast waves with w in homogeneous unstratified coronal loop in the presence of thermal conduction.

acoustic waves modified by the presence of the magnetic field, while fast waves can be seen as Alfvén waves modified by the compressibility of the plasma. Under the same considerations, the frequency of waves can be approximated by $\omega^2 \approx k_z^2 c_S^2$ for slow waves and $\omega^2 \approx K^2 v_A^2$ for fast waves. In order to study the dependence of the damping length of these modes we suppose a real frequency and a complex longitudinal wavenumber, i.e. $k_z = k_r + ik_i$. The wavelength of waves is given by $\lambda = 2\pi/k_r$, while the damping length by $L_d = 1/|k_i|$. We also suppose that waves will have a weak damping, expressed mathematically by the inequality $|k_r| \gg |k_i|$. If this form of the wavenumber is inserted back into Eq. (12) from, then we can isolate the imaginary part of the wavenumber giving us the damping length of waves. Solved for slow waves, the damping length will take the simple form

$$L_d^{slow} \approx \frac{3w^2 c_S(\gamma-1)}{\chi k_r^2 (w^2-\gamma+1)}, \tag{13}$$

where $w = k_\perp/k_r$ is the ratio of the transversal and parallel wave number. We should note that this damping length does not depend on the presence of the magnetic field but it does depend on the transversal wavenumber. By requiring that slow waves have a damping ($k_i < 0$), we obtain that the condition for damped slow waves is $k_\perp/k_r > \sqrt{\gamma-1}$. Given the temperature dependence of the sound speed and the coefficient χ we obtain that the damping length is proportional to $T_0^{-1/2}$.

>From the same physical considerations, the damping length in the case of fast magnetoacoustic waves is given by

$$L_d^{fast} = 2\frac{(\gamma-1)\, v_A\, c_S{}^2\sqrt{w^2+1}}{k_r{}^2\chi\, [(\gamma-1)\,(w^2+1)\, v_A{}^2 - w^2 c_S{}^2]}. \tag{14}$$

The variation of the damping lengths for slow and fast waves with respect to the dimensionless quantity w is shown in Figure 1.

This numerical analysis show that the damping length of slow waves are roughly the double the "detection length" observed by De Moortel et al. (2004).

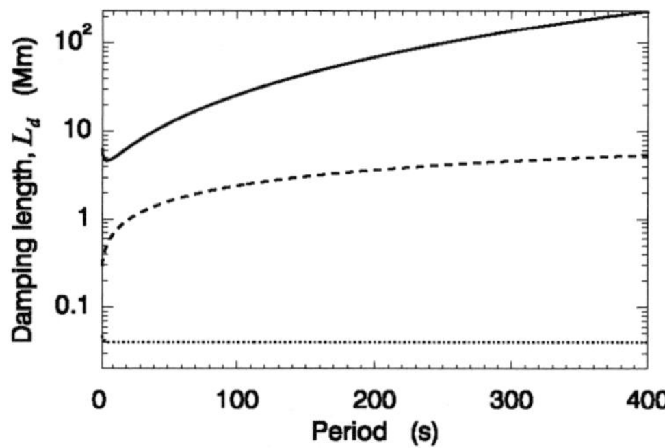

FIGURE 2. The variation of the damping length with respect to the period for slow (dashed line),fast (solid line) and thermal (dotted line) waves.

If we solve numerically the dispersion relation (12), written for z component of the wavenumber,it is possible to illustrate the variation of the damping length with respect to the period of waves for slow, fast and thermal mode (see Figure 2).

This figure shows that for periods between 2 and 400 seconds, the damping length of the thermal modes is practically independent on the period and it has a very small value, i.e. this mode will damp very quickly. For the considered period range the damping length of slow and fast waves increases with increasing the period. In the case of fast magnetoacoustic modes the damping length has a minimum of 3 Mm for periods of the order of 5 seconds. After this value the damping length increases such that for periods longer than 200 seconds it reaches lengths of over 100 Mm. In the case of slow waves, the damping length has a slight increase except for periods shorter than 40 seconds where this increase is more abrupt. The remarkable property of the damping length of slow waves is that it remains relative low even for very long periods.

The effect of the gravity can be best shown on a graph which plots the relative change of the damping length compared to the case with no gravity. To calculate this quantity we used the relations derived in the previous subsection together with Eq. (12) and in Figure 3 we plot the quantity $1-L_d/L_d^{gravit}$ on a logarithmic scale where L_d is the value derived in the case of a homogeneous unstratified plasma while L_d^{gravit} is the damping length in the presence of gravity.

In the case of the fast magnetoacoustic modes, the damping length is considerable increased by the presence of the gravity for the entire considered temporal domain. In the case of slow waves, the inclusion of gravity increases the damping length for periods up to approximately 30 seconds. For waves with periods larger than this value, the damping length is decreased (a similar result was obtained for hot SUMER loops by Mendoza-Briceno et at. 2004). The peak in the variation of the damping length for slow waves with respect to the period of waves appears simply because for that value of period, the damping lengths corresponding to homogeneous and gravitationally stratified plasma are very close. In the case of thermal modes, the gravity will introduce a small decrease for the entire domain of periods.

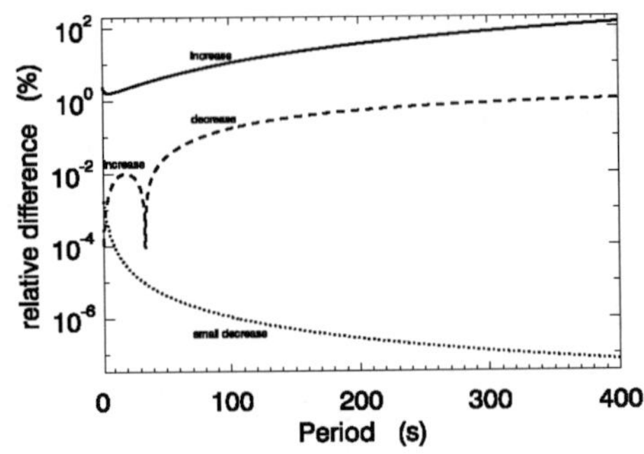

FIGURE 3. The relative variation of the damping length with respect to the period for fast (solid line), slow (dashed line) and thermal (dotted) waves in the presence of gravity.

CONCLUSION

The presence of an additional transversal motion has changed considerable the result obtained previously by De Moortel and Hood (2003, 2004).We have shown that the damping length of slow MHD waves is comparable to the damping lengths observed by De Moortel et al. (2002). In the case of fast MHD waves, thermal conduction leads to a weak damping. The gravitational stratification changes slightly the damping length of slow MHD waves, while in the case of fast waves, this increase is considerable.

ACKNOWLEDGMENTS

The authors would like to acknowledge the financial support received from the organizers of the meeting. I.B. acknowledges the support offered by the Nuffield Foundation (NUF-NAL 04) and NSF Hungary (OTKA, TO43741).

REFERENCES

1. L. Ofman, M. Romoli, G. Poletto, C. Noci, J. L. Kohl, *ApJ* **491**, L111 (1997).
2. C. E. DeForest and J. B. Gurman, *ApJ* **501**, L217 (1998).
3. D. Berghmans and F. Clette, *Sol. Phys.* **186**, 207 (1999).
4. I. De Moortel, and A. W. Hood, *A&A* **408**, 755 (2003).
5. I. De Moortel, and A. W. Hood, *A&A* **415**, 705 (2004).
6. B. J. Thompson et al., *ApJ* **517**, L151 (1999).
7. I. Ballai, R. Erdélyi, B. Pintér, *ApJ* **633**, L145 (2005).
8. V. M. Nakariakov, L. Ofman, E. Deluca, B. Roberts, J. M. Davila, *Science***285**, 862 (1999).
9. I. Ballai, *Sol. Phys.*, in press (2007).
10. P. Edwin and B. Roberts, *Sol. Phys.* **88**, 179 (1983).
11. M. S. Ruderman, R. Oliver, R. Erdélyi, J. L Ballester, M. Goossens, *A&A* **354**, 261 (2000).
12. S. I. Braginskii, *Rev. Plasma Phys.* **1**, 205 (1965).
13. I. De Moortel, A. W. Hood, J. Ireland, R. W Walsh, *Sol.Phys.* **209**, 89 (2002).
14. I. Ballai, A. Marcu, B. Pintér, *A&A*, submited (2007).
15. C. A Mendoza-Briceno, R. Erdélyi, L. Di L Sigalotti, *ApJ* **605**, 493(2004).

Analysis of Four Coronal Mass Ejections using SOHO Data

M. Mierla* and S. Giordano†

*Astronomical Institute of Romanian Academy, Bucharest, Romania
†INAF-Astronomical Observatory of Torino, Pino Torinese, Italy

Abstract. By using the data from different instruments of SOHO we can determine the properties of four back side coronal mass ejections from inner regions (MDI, EIT and C1 images) up to around 30 solar radii (C3 images). From MDI back side images the source of coronal mass ejections can be inferred. From UVCS and LASCO-C1 spectral data, plasma speeds and temperatures can be determined. The properties of higher altitudes of the solar corona will be deduced from LASCO-C2 and -C3 white light images.

Keywords: Sun, corona, spectral lines, coronal mass ejections
PACS: 90.60.Q

INTRODUCTION

Coronal mass ejections (CMEs) are enormous eruptions of plasma ejected from the Sun into interplanetary space, over the course of minutes to hours. The first CMEs were observed in 1970s by coronagraphs aboard the Orbiting Solar Observatory and Skylab. Since then, many spatial missions have recorded thousands of such events, the most recent, and still working, being Large Angle Spectroscopic Coronagraph (LASCO), onboard the Solar and Heliospheric Observatory (SOHO) mission. The rate of occurrence of CMEs changes along the solar cycle, with an average value of around 1 event every 2 days during solar minimum and around 4 events during solar maximum (based on the number of SOHO CMEs per year; [2]). Average speeds of CMEs are of the order of 400 km s^{-1} ([1]; [2]), although they may exceed 2000 km s^{-1}. The angular spans may extend from just a few degrees in the case of narrow CMEs, up to 360 for "halo" CMEs that completely surround the occulter of the coronagraph. A high percentage of CMEs shows a pronounced three-part structure. These events appear as having a bright leading edge, a dark void and a bright core ([3]; [4]). In this paper we analyze four CMEs which occur on March 1998, from the same region of the solar disk. Different SOHO instruments are used for this analysis.

THE INSTRUMENTS AND OBSERVATIONS

The Michelson Doppler Imager (MDI) is the instrument of Solar Oscillations Investigation (SOI) to investigate the solar interior by measuring the photospheric manifestations of solar oscillations ([5]). It provides high resolution images of the line of sight (LOS) velocity, line intensity, continuum intensity, longitudinal magnetic field components and

CP934, *Flows, Boundaries, Interactions*
edited by C. Dumitrache, V. Mioc, and N. A. Popescu

limb position, by means of sampling the Ni I 676.8 nm line with a wide-field tunable Michelson Interferometer. The Extreme-Ultraviolet Imaging Telescope (EIT) observes the solar corona and transition region, on the solar disk and up to a distance of 1.5 $R_\odot$. It records these regions in 4 spectral lines: He II 30.3 nm (80000 K), Fe IX 17.1 nm (1.1 MK), Fe XII 19.5 nm (1.5 MK) and Fe XV 28.4 nm (2 MK) ([6]). The Large Angle and Spectrometric Coronagraph (LASCO) ([7]) consists of three individual coronagraphs with nested fields of view. The C1 coronagraph observes the solar corona from 1.1 to 3.0 $R_\odot$ and contains a Fabry-Perot (FP) interferometer that allows imaging the corona in different emission lines. The C2 and C3 coronagraphs records the white-light corona from 2 to 6 $R_\odot$ and 3.7 to 30 $R_\odot$, respectively ([7]). The Ultraviolet Coronagraph Spectrometer (UVCS) observes the solar corona between about 1.4 and 10 $R_\odot$ ([8]) with an instantaneous field of view of 40 arcmin perpendicular to the sunward direction on the plane of the sky. The entrance slit can be rotated by 360° around an axis pointing to the center of the Sun to cover all the possible position angles. It has two ultraviolet spectral channels. The O VI channel covers the range 94.5-112.3 nm (47.3-56.1 nm in second order), and the LYA channel covers 116.0-127.0 nm.

The data analyzed here are images taken by the above mentioned instruments on March 1998. The MDI data are whole-Sun maps of magnetic regions. The Earthside, or near-side, data are smoothed magnetic flux as measured by MDI. The farside images are maps of wave speed variations with locations of faster wave speed shown darker. These darker regions indicate locations where there is an accumulation of magnetic field on the far surface. The LASCO-C1 data are images taken in the lines of Fe XIV at 530.3 nm (green coronal line). They consist of series of about 47 sets (from 28th to 30th of March) each consisting of two scans of both the west and east limb of the Sun. Each scan contains 15 images at wavelengths ranging from 530.20 nm to 530.65 nm in steps of 0.03 nm, along with 2 off-line images at wavelengths of 531.14 nm and 529.99 nm (wavelengths are given relative to vacuum). The images were taken with a cadence of 1 min, the exposure time being 25 s. Thus, spectra are recorded in around 15 min ([9]). The LASCO-C2 and -C3 data consists of white-light images which map the visible photospheric light scattered by the free electrons in the solar corona. The UVCS data consists of a series of two-dimensional images, where the axis are the coordinate along the slit and the wavelength dispersion direction. The main spectral lines detected are the H I Lyα at 121.6 nm, and the O VI doublet lines at 103.19 nm, and 103.76 nm. The UVCS slit has to be placed at the right time and location in the corona to observe spectra of a CME as it crosses the field of view of the slit.

DATA ANALYSIS AND INTERPRETATION

Analyzed Events Overview

The four CMEs (Fig. 1) we analyze here are coming from a bright complex active region (AR) structure: AR 8176, AR 8178 and AR 8179. The first one, on 23rd of March, 09:33 UT, PA 271 (Fig. 1, upper left panel), was analyzed by [13] and catalogued as a back side event by [10]. The next 2 CMEs, took place on 25th of March (13:14 UT, PA 242) (Fig. 1, upper right panel) and 27th of March (20:07, PA 209) (Fig. 1, lower left panel), respectively. They were structured CMEs (see for e.g. [10], for definition of structured CMEs). The last one, on 29th of March (Fig. 1, lower right panel), was a halo CME.

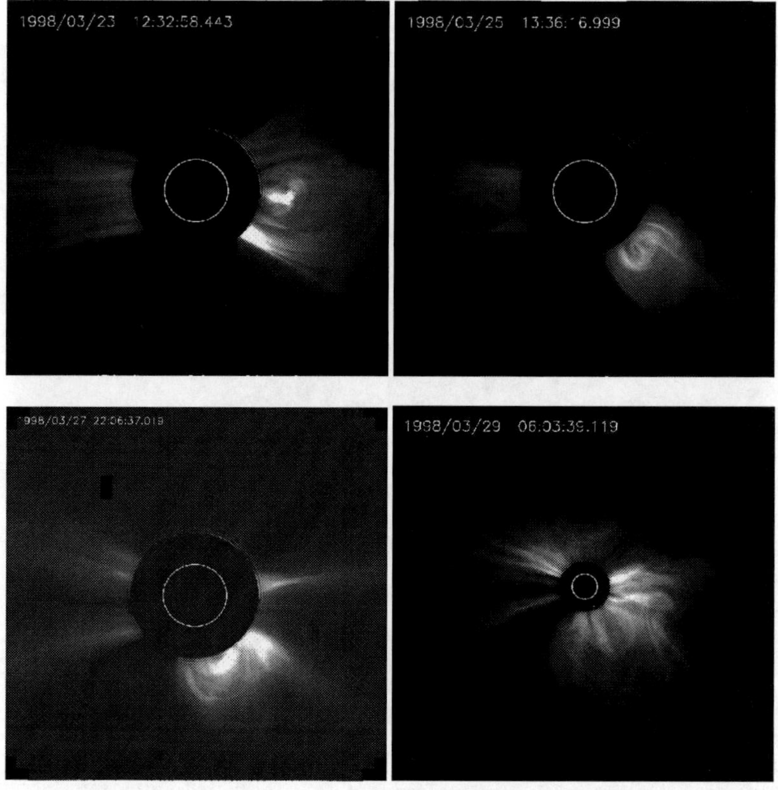

FIGURE 1. LASCO-C2 white-light images recorded on 23rd of March 1998, 12:32UT (upper left panel), 25th of March 1998, 13:36UT(upper right panel), 27th of March 1998, 23:30UT (lower left panel) and LASCO-C3 image recorded on 29th of March, 06:03UT (lower right panel).

The changes in polar angle (see Tab. 1) also indicate that the four CMEs are coming from the same region on the solar disk (the source region changes as the Sun rotates).

TABLE 1. CMEs parameters from LASCO CME catalogue (cdaw.gsfc.nasa.gov/CME_list). The date and time are given by the first C2 appearance, PA is the central position angle in degrees, measured counterclockwise from the north pole, the speed in km/s are from the height-time measurements of the leading edge.

Date	Time (UT)	PA	Speed
23.03.98	09:33	271	200km/s at 2.5Rs - 600km/s at 28Rs
25.03.98	13:14	242	761km/s from 4Rs to 28Rs
27.03.98	20:07	209	450km/s at 3Rs - 600km/s at 26Rs
29.03.98	03:48	halo	1400km/s from 9Rs to 24Rs

The CME on 23rd of March

AR 8179 is the biggest and most complex one, giving rise to a series of flares and CMEs. As seen in Fig. 2, on 20th of March, it is almost at the west limb and its neutral line is parallel with the solar limb. On 23rd of March the AR is on the back side of the disk and its neutral line is perpendicular to the solar limb. As discussed by [10], this kind of geometry gives rise to structured CMEs, as seen in Fig. 1. They consist of a bright core, a dark cavity and a bright leading edge. A detailed description of the CME is given in Ciaravella et al. 2002. Using UVCS data, they could detect the reconnection current sheet, as predicted by the model of Lin and Forbes 2000.

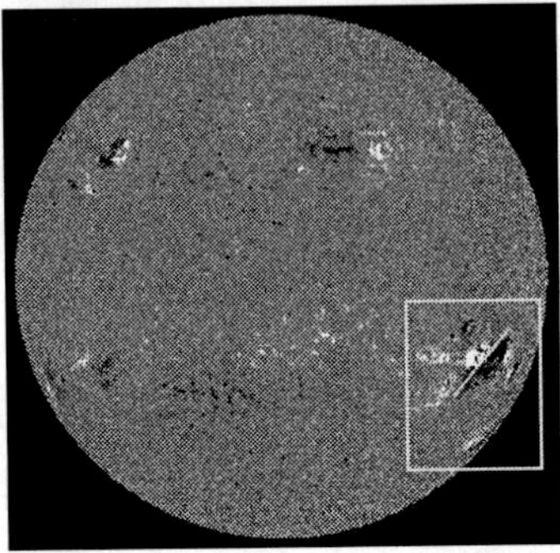

FIGURE 2. MDI magnetogram recorded on 20th of March 1998, 01:36 UT

The CME on 25th of March

A structured CME is observed in C2 field of view, on 25th of March, starting from 13:14UT, centered in the south-west quadrant at 242° with an angular width of 104° (see Fig. 1 upper right panel). At that time UVCS was performing the south pole scan of the synoptic program. In particular UVCS observed at 2.36 $R_\odot$ from 12:44UT to 13:29UT and at 2.64 $R_\odot$ from 13:31UT to 14:31UT, with 300 seconds temporal resolution, covering a latitudinal range from 162° to 212°. Therefore UVCS was able to observe the south leg of the CME. The intensity started to increase, both in H I Lyα and O VI 1032, at 12:59UT when the UVCS slit was pointed at 2.26 $R_\odot$, in agreement with LASCO-C2 data. The brightest emission is detected in the region from 207° to 203° slightly drifting in time toward south pole (as expected by the CME expansion in the tangential direction). The UVCS data provide a temporal scan of the event at a fixed position. With the assumption that the gas moved at constant speed of 761 km/s (see Tab 1) and the approximation of a radial expansion we can reconstruct a two-dimensional image of the observed region. Each 300 seconds exposure takes the spectrum from the coronal region observed at each spatial bin (21 arcseconds wide) in the instantaneous field of view. The integration of a selected spectral line over its profile provides the intensity distribution along the slit at that particular time. By placing the intensities from consecutive exposures along a time-reversed axis, and radially expanding the covered

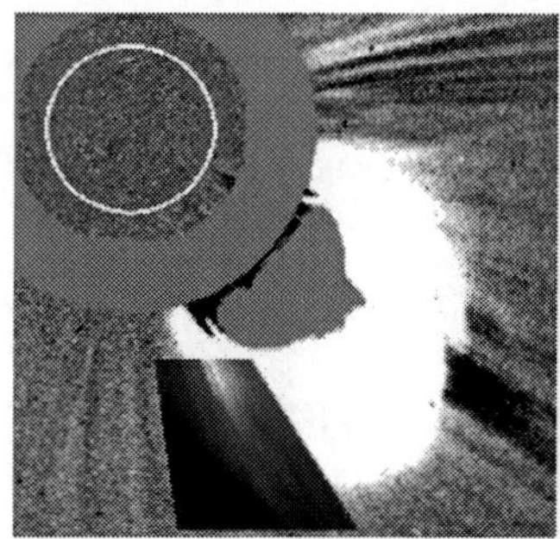 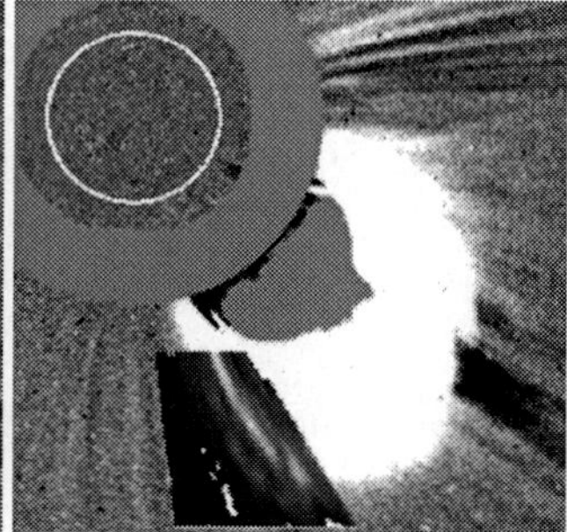 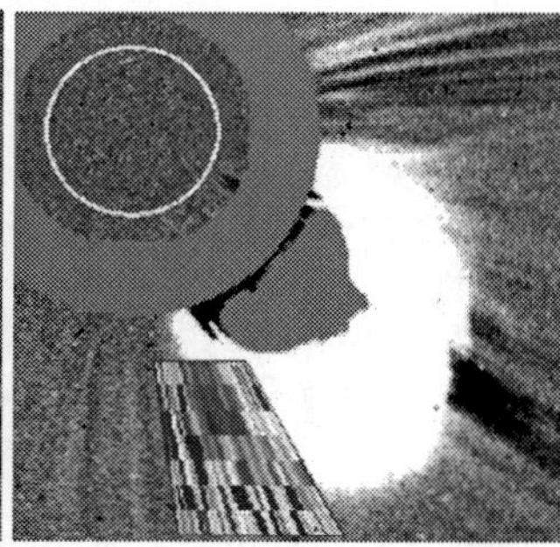

FIGURE 3. LASCO-C2 difference image taken on 1998 March 25 at at 13:36UT with superimposed the reconstructed UVCS intensity images in H I Lyα (left panel) and O VI 1032 (central panel) taken from 12:44UT to 13:29UT. In the right panel is superimposed the H I Lyα Doppler shift, the red and blue colors represent the red and blue shift, respectively. The bright leg is predominantly red-shifted, while the blue-shift is manly due to noise.

region as a function of time, we obtain the two-dimensional images which corresponds to the images at the time of the last spectrum. Fig. 3 show the reconstructed intensity images of H I Lyα and O VI 1032, from data at 2.36 $R_\odot$, superimposed to the LASCO-C2 difference image taken at 13:36UT. The images show the same morphological structure of the south CME leg as the LASCO-C2 white-light images (see also Fig. 1 upper right panel). From H I Lyα spectral data we compute the line centroids to evaluate the Doppler shifts with respect to the exposures taken before the CME crosses the slit, then as for the intensity we build a two-dimensional image. The H I Lyα Doppler shift superimposed to LASCO-C2 image (right panel of Fig. 3) shows a predominant red-shift in the region where the bright leg is observed. A detailed analysis of the line centroids shows a red-shift up to about 100 km/s, showing that it is a back side CME.

The CME on 27th of March

On 27th of March, at around 20:07 UT another structured CME is observed in C2 field of view (Fig. 1, lower left panel). The MDI far side images show how the magnetic field

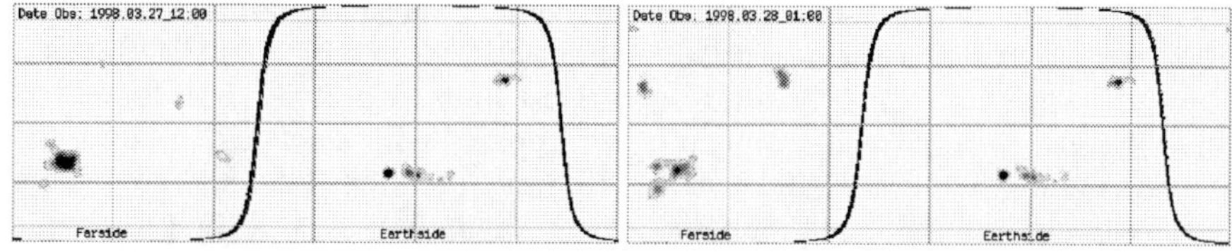

FIGURE 4. MDI farside and Earthside images taken on 27th of March 1998, 12:00UT (left panel) and on 28th of March, 00:00UT (right panel).

intensity changes in this period. We have a strong magnetic field before the ejection and a weaker one, after (Fig. 4). Also, after the event, the AR looks like separated in three regions. No activity is visible on the front side of the disk.

The halo CME on 29 March

On 29th of March we observe in LASCO-C2 data a halo CME (Fig. 1, lower right panel) beginning at 03:48 UT. The front speed of this CME was almost 1400 km/s, qualifying it as an extremely fast event. The CME front edge passes 2 $R_{\odot}$ at around 03:00. In this period of time we do not have C1 spectral data. Still, the signature of the plasma following the CME could be observed in the spectral data by the strong red shift on 29th of March, 04:21 UT (third row, middle image, west limb of Fig. 5), long after the CME has passed the field of view of C1. This red shift give us indication that it was a back side event ([12]). From the FWHM maps we see that the temperature did not change significantly during the plasma passage compared with the previous conditions.

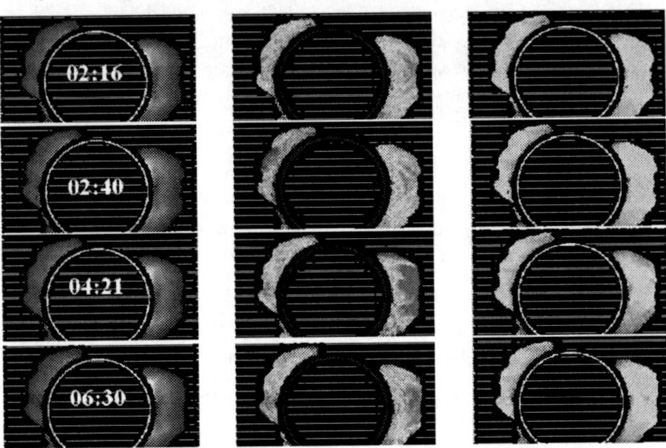

FIGURE 5. LASCO-C1 emission maps taken on 29th of March 1998. The left panels represent the radiance maps, the middle ones: the speed maps and the right ones: the full width at half maximum (temperatures) maps. The time varies from up down as following: 02:16UT, 02:40UT, 04:21UT and 06:30UT.

DISCUSSION AND CONCLUSIONS

The AR complex we have described here was analyzed over one and a half solar rotation by [13] (UVCS data) on 20-24 March 1998 at south-west limb, [14] (UVCS data), half rotation later: on 1998 April 6-7 at the south-east limb, and by [?] (CDS data), half rotation later: on 1998 April 20 at the south-west limb . Ko et al. 2002 found that the absolute abundances are smaller in their data compared with the data of Ciaravella et al. 2002, showing that while the AR evolves with time, the loops become more static and the materials stored within are gradually settled gravitationally. [?] (half rotation later) found smaller temperatures as compared with the temperatures found by Ko et al. 2002, showing that the AR cooled in time. We have analyzed this AR complex between 23rd of March and 29th of March, period when it was on the back side of the solar disk. Four CMEs originated in this AR in the above mentioned period. The CMEs on 23th, 25th and 27th of March were structured CMEs and that of 29th of March was a halo CME. Their signatures can be seen on Carrington C2 map (Fig. 6) as vertical bright stripes.

The streamer seen on south-west limb, appears on 19th of March. It stays there up to

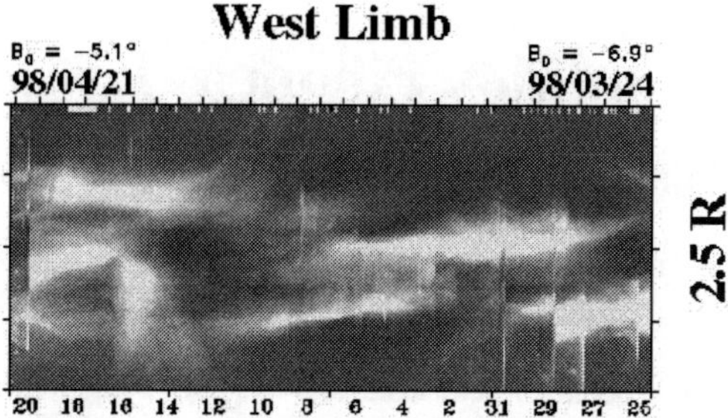

FIGURE 6. LASCO-C2 Carrington map of the west limb showing the large-scale solar corona during 27 days (1 rotation). The time runs from right to left (courtesy SOHO/LASCO).

29th of March when it is blow up by the halo CME. The other 3 CMEs just displaced it from its position in the period they occurred. The CME on 25th of March was observed also by UVCS instrument. From the Doppler analysis we could see that it was a back side CME, having speeds on the line-of-sight at around 100 km s^{-1}. The halo CME on 29th of March was observed with LASCO-C1 spectral data, showing a strong red shift in the speed maps. The AR from where the four CMEs initiated could be observed in MDI farside images.

ACKNOWLEDGMENTS

We would like to thank the SOHO consortium for providing the data and the software libraries. The CME catalog is generated and maintained at the CDAW Data Center by NASA and The Catholic University of America in cooperation with the Naval Research Laboratory. SOHO is a project of international cooperation between ESA and NASA.

REFERENCES

1. A. J. Hundhausen, J. T. Burkepile, O. C. St.Cyr *J. Geophys. Res.* **99**, 6543–6552 (1994).
2. S. Yashiro, N. Gopalswamy, G. Michalek, et al., *J. Geophys. Res.* **109**, 7105– (2004).
3. R. M. E. Illing, A. J. Hundhausen *J. Geophys. Res.* **90**, 275–282 (1985).
4. R. A. Howard, N. R. Sheeley, D. J. Michels, M. J. Koomen *J. Geophys. Res.* **90**, 8173–8191 (1985).
5. P. H. Scherrer, R. S. Bogart, R. I. Bush, et al. *Solar Phys.* **162**, 129–188 (1995).
6. B. J. Thompson, S. P. Plunkett, J. B. Gurman, et al. *Geophys. Res. Lett.* **25**, 2465–2468 (1998).
7. G. E. Brueckner, R. A. Howard, M. J. Koomen, et al., *Solar Phys.* **162**, 357–402 (1995).
8. J. L. Kohl, R. Esser, L. D. Gardner, et al., *Solar Phys.* **162**, 313–356 (1995).
9. M. Mierla, R. Schwenn, L. Teriaca, G. Stenborg, B. Podlipnik, *Adv. Space Res.* **35**, 2199–2203 (2005)
13. A. Ciaravella, J. C. Raymond, J. Li, et al. *Astroph. Journal* **575**, 1116–1130 (2002).
10. H. Cremades, V. Bothmer *A&A* **422**, 307–322 (2004).
11. J. Lin, T. G. Forbes *J. Geophys. Res.* **105**, 2375–2392 (2000).
12. Mierla M., "On the Dynamics of the Solar Corona", *PhD thesis*, Copernicus GmbH, Goettingen (2005)
14. Y. K. Ko, J. C. Raymond, J. Li, et al. *Astroph. Journal* **578**, 979–995 (2002).
15. J. T. Schmelz, R. T. Scopes, J. W. Cirtain, et al. *Astroph. Journal* **556**, 896–904 (2001).

Huge Complex Filament Evolution

D.R. Constantin, M. Mierla, C. Oprea and C. Dumitrache

Astronomical Institute of the Romanian Academy, Bucharest, Romania, e-mail: diana@aira.astro.ro

Abstract. A complex filament composed by a main body(a huge polar filament), and a tail (a small filament situated between active regions) was observed between 6 and 14 January 2001. We have analyzed its dynamics. The tail suffered a thermal sudden disappearance and reappeared after two days. A decaying active region plays the role of attractor for the complex filament. A spectacular CME, produced after a helical up-awarded movement of plasma, movement observed in the coronal loops of the main body of the filament during three days before the CME onset. We have analyzed also the CME shape and movement using SOHO data.

Keywords: prominence, H alpha, chromosphere, corona
PACS: 96.60.qf, 96.60.Na, 96.60.P-

INTRODUCTION

The prominences are cold features of the hot solar corona, and they are anchored in the subphotospheric layers of the Sun. When they are observed on the solar disk, they are known as filaments. The prominence channels form along the polarity inversion lines, which make from prominences veritable tracers of the solar magnetic field. These lines appear inside active regions, either between active regions, or between the active regions and the unipolar magnetic field. The quiescent prominences often could attain big dimensions. The polar crown filaments become often huge, covering almost 40 degrees of the Sun's disk, especially in the neighborhood of the reversal of solar polar magnetic field. The Northern hemisphere registered the reversal of the magnetic field polarity in April 2001 [1].

A complex and huge filament was observed between 6 and 14 January 2001. We analyzed this filament using the following data: Hα full disk observations from Big Bear Solar Observatory (BBSO) and also from Meudon Observatory, the magnetograms from MDI/SOHO and coronal observations of EIT/SOHO, as well as the data of C2/SOHO. The filament is situated in the Northern hemisphere of the Sun (Figure 1), during the Carrington rotation No. 1971, as displayed also on the "chartes synoptiques" Meudon.

COMPLEX FILAMENT EVOLUTION

The filament has two components, the main body ($f1$), that is a polar filament, and the tail ($f2$) situated among three active regions - AR09294, AR09296 and AR09299. Figure 1 displays the Hα images of the filament on 6, 7, 8, and 9 January 2001, while Figure 2 displays these images on 10 and 11 January. The AR09299, which appeared on 4 January, disappeared on 8 January. On 8 and 9 January the tail disappeared, too,

CP934, *Flows, Boundaries, Interactions*
edited by C. Dumitrache, V. Mioc, and N. A. Popescu

but on 10 January the tail reappeared and was observed until the filament arrived at the solar border. In this period no CME or flare were registered in the region, so we suspect that the tail supported a thermal disappearance on 8 and 9 January. These kind of disappearances were described by [2]. Notice that a surge occurred in the tail on 8 January, between 07h13m and almost 23h (Figure 3).

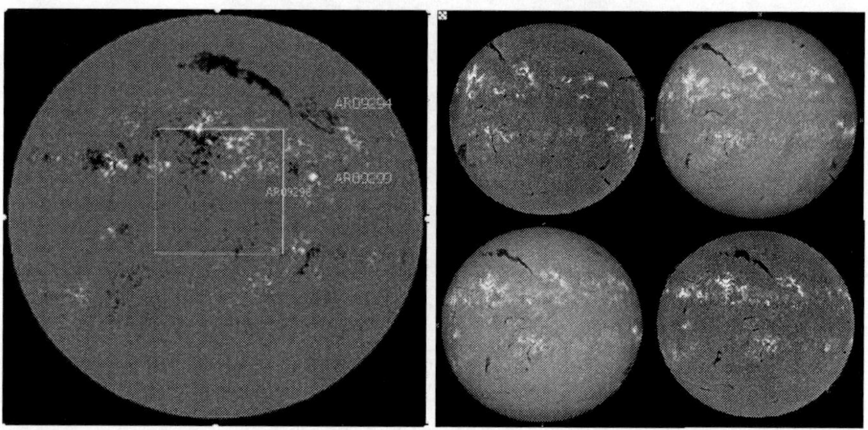

FIGURE 1. Hα filament superposed on MDI magnetogram (left panel); Hα filament on 6, 7, 8 and 9 January 2001 (right panel).

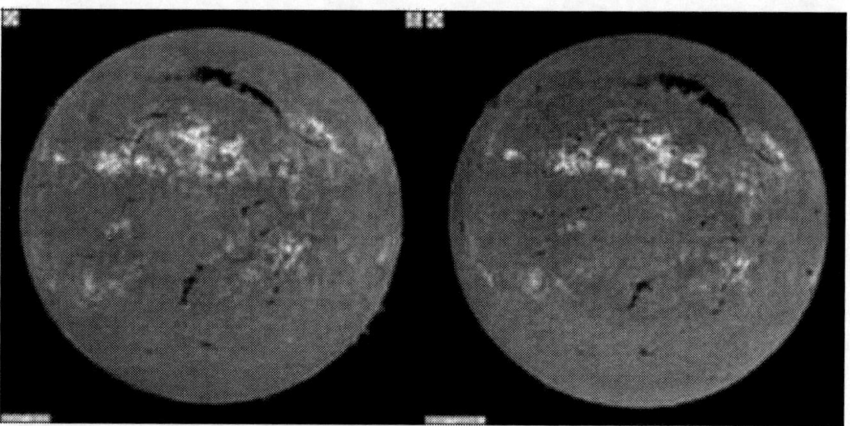

FIGURE 2. Hα filament on 10 and 11 January 2001.

The main body of this huge filament and its tail had a spectacular dynamics during its transit on the solar disk (between 6 and 14 January 2001), changing not only its shape and thickness, but also orientation.

In order to analyze the complex filament (*f1+f2*) evolution, we have computed its length and inclination on the solar parallel as follows [3]:

$$LL = \arccos(\sin\varphi 1 \cdot \sin\varphi 2 + \cos\varphi 1 \cdot \cos\varphi 2 \cdot \cos(L1 - L2)) \quad (1)$$

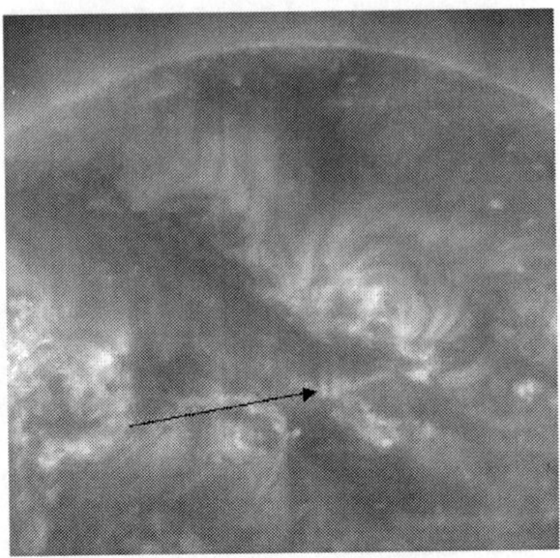

FIGURE 3. Surge observed in SOHO/EIT on 8 January 2001.

$$u = \arcsin\left(\frac{\sin\varphi 1 - \sin\varphi 2 \cdot \cos LL}{\cos\varphi 2 \cdot \sin LL}\right) \tag{2}$$

where $(\varphi 1, L1)$ and $(\varphi 2, L2)$ are the coordinates of the ends of the filament, $L1 > L2$. LL represents the length of the filament computed as an arc of great circle on a sphere, whereas u is the angle made by the filament with the solar parallel. For $u < 0$ the filament is dextral, while for $u > 0$ is sinistral.

The length of the filament is function of its inclination on the solar parallel, its differential rotation and latitude, and time (day of the year), $LL = LL(u, w, \varphi, t)$, with $\varphi = \frac{\varphi 1 + \varphi 2}{2}$, i.e., the computed length of the filament is

$$LLc = pwt\sqrt{\left|\frac{\varphi}{u}\right|} \tag{3}$$

where p is a fitting parameter [4]. The differential rotation law $w = |\omega(\varphi)|$, where $\omega(\varphi)$ is the well known d'Azambuja law [5]:

$$\omega(\varphi) = 1.12 - 1.4\sin^2\varphi - 11.33\sin^4\varphi \tag{4}$$

Figure 4 displays the graphics of the *f1* length, *f2* length and the length of the complex compound by *f1*+*f1*. A plot of the tilt angle dynamics and also for the differential rotation variation could be seen for each day in Figure 5.

On these graphics we can survey the dynamics of both filaments. On 9 January, the length of the main body of filament had a maximum value in spite of the tail has disappeared; this explains why in same day we have a maximum of the whole filament's length. On Hα images one observes that the tail's length diminished next days, especially on 11 January. From 10 to 13 January, the distance between the main body *f1* and its tail increased and this explains why the plot of whole filament's length has a maximum value on 11 January, when the tail's length had a minimum value in the

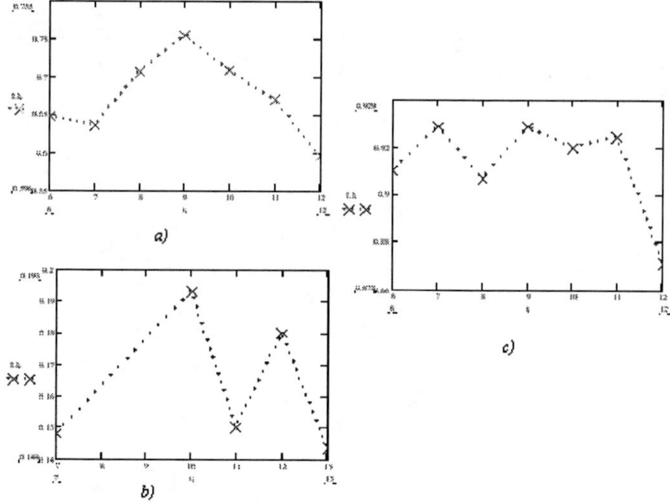

FIGURE 4. (a) The evolution of the main body of filament length; (b) The evolution of the tail length; (c) The evolution of whole filament length.

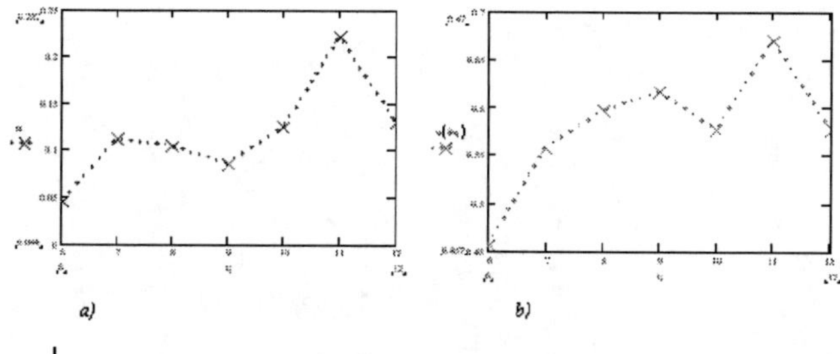

FIGURE 5. (a) The tilt angle dynamics; (b) The differential rotation variation.

same day. On 11 January we found out a maximum value for the differential rotation, and also for the absolute value of the tilt angle.

The decaying AR 09299 plays a role of an attractor (sink) for both parts of the filament (the main body and the tail). The tail went away Southward from the main. On 11, 12,1 3 January one could observe flows inside the main body *f1*. These flows are oriented towards the tail and AR09299 direction, giving the illusion that the filament is brushed in that direction (Figure 6). These motions are more clearly seen on the EIT films. A helical up-warded motion observed by EIT (195 Å) on 11, 12, 13, and 14 January 2001 occurred and the loops are up-lifted (Figure 5, right panel). Beginning with 11 January the loops were continuously uplifted to erupt on 14 January 2001, when a spectacular CME produced. The CME development observed on EIT wavelet images is displayed in

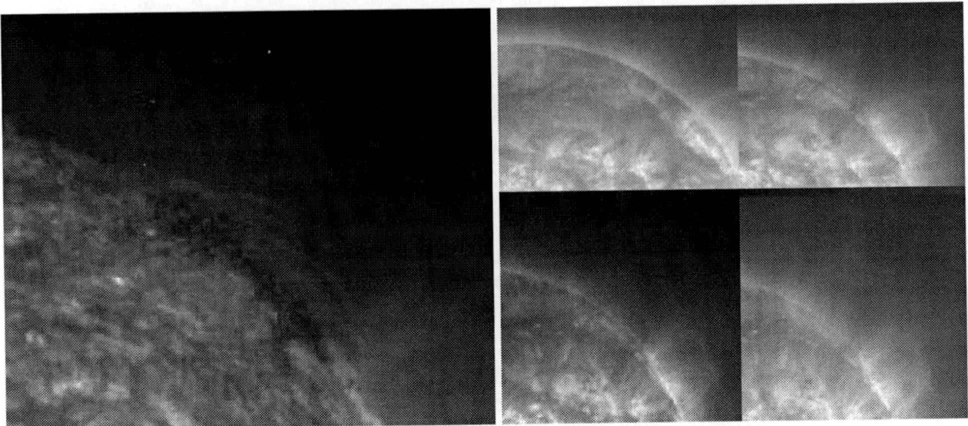

FIGURE 6. The SOHO/EIT (304 Å) image on 11 January 2001(left panel); a helical up-warded movement observed by EIT (195 Å) on 11, 12, 13 and 14 January 2001(right panel).

Figure 7.

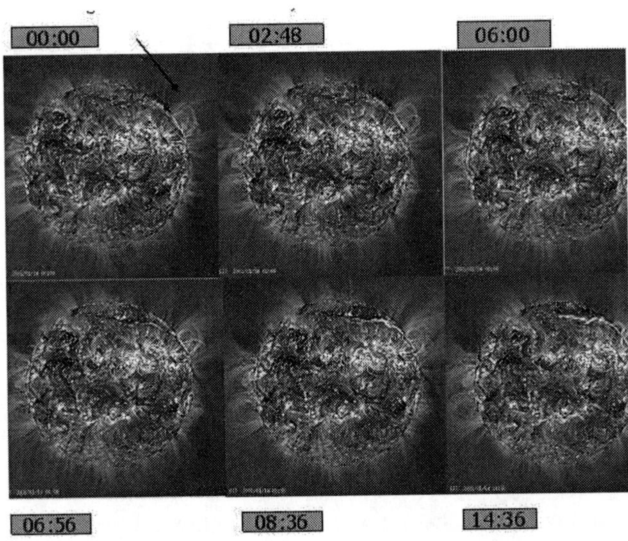

FIGURE 7. The EIT wavelet images of the CME at different moments on 14 January 2001.

At the beginning of CME, $f1$ erupted in many phases: first erupted the loops arrived at a high helicity. This area is near AR09299, disappeared on 8 January, and played the role of an attractor (sink) for the plasma of the whole filament. The first eruption of the filament (at about 3h12m) triggered the whole filament. At 6h36m a second point of eruption occurred in the Northern part of the filament. The CME extends also there and contained the whole filament and its Northern neighborhood on an extended area. This

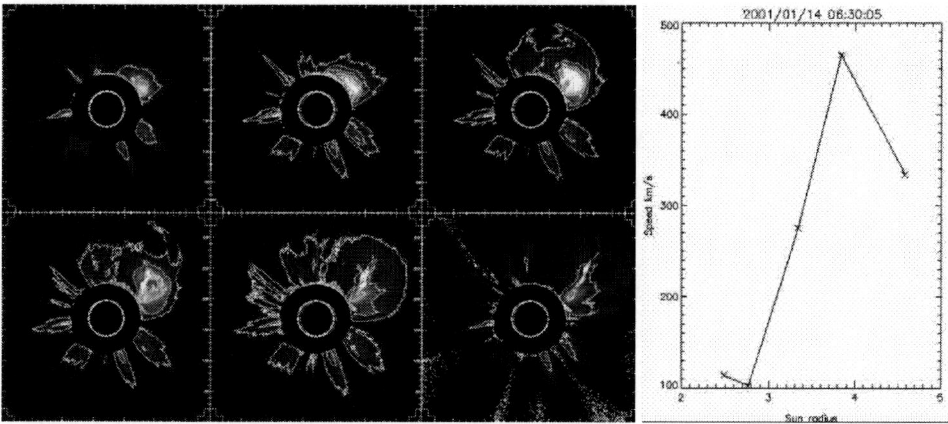

FIGURE 8. The CME observed by LASCO/C2 (left panel); The speed plot on 14 January 2001 (right panel).

second point could be seen on LASCO/C2 registrations occurring at about 7h30m. We plotted them together with the contours plots to emphasize the CME shape in Figure 8. We have no information about the tail behavior during this CME.

Figure 8 (right panel) displays the characteristics of the blob movement originating from the first point of eruption. We followed this blob between 2 and 5 solar radii. We observed an acceleration until 4 solar radii, and a deceleration near 5 solar radii. We have not the next point (behind 5 solar radii) to give a pertinent explanation of this fact. If it is not an error of evaluation (the last point was at the border of our image), the deceleration could be a real phenomenon, similar to the one studied by [6].

CONCLUSIONS

We have studied a complex structure composed by a main body (polar filament) and a tail (formed between the active regions AR09294, AR09296 and AR09299).

The dynamics of the tail was influenced by the disappearance of an active region - magnetic reconnections that lead to a thermal sudden disappearance.

The main body dynamics displays a high helicity and this is explained by the fact that AR09299 disappearance plays a role of attractor for the complex filament, especially for the polar filament. This phenomenon could be explained by the action of Lorentz force.

On EIT images one can observe, better than in Hα, that the main body of the filament elongated Northwards after 12 January, disparaging any other small filaments structure in its way on the solar parallel. On that day, when the filament arrived at the solar border, it became a huge object. This "cannibalism" behavior produced an easy fragmentation at the middle of the main body.

An uplift plasma motion in the main body of the filament ended in a CME on 14 January 2001. The main body of the filament is influenced by many actions, but in principal there are two mechanisms: its Southern part attraction by the disappeared active region, and its "cannibalism" behavior in the Northern part, where a second point of CME occurred, probably as a consequence of a new flux emergence.

ACKNOWLEDGMENTS

We would like to thank the SOHO consortium for providing the data and the software libraries. The CME catalog is generated and maintained at the CDAW Data Center by NASA and the Catholic University of America in cooperation with the Naval Research Laboratory. SOHO is a project of international cooperation between ESA and NASA.

REFERENCES

1. E. E. Benevolenskaya, *A&A* **428**, L5–L8 (2004).
2. Z. Mouradian, I. Soru-Escaut, *Hvar Obs. Bull.* **13**, 379–391 (1989).
3. C. Dumitrache, *Solar Phys.* **173**, 281–304 (1997).
4. C. Dumitrache, C. Oprea, D. Constantin, M. Mierla, A. S. Popescu, in C. Dumitrache, N. A. Popescu, D. M. Suran, V. Mioc (eds), *Fifty Years of Romanian Astrophysics, AIP Conf. Proc.* **895**, 69–74 (2007).
5. L. d'Azambuja, M. d'Azambuja, *Ann. Obs. Paris* **6**, fasc. 7 (1948).
6. D. Tripathi, S. K. Solanki, R. Schwenn, V. Bothmer, M. Mierla, G. Stenborg, *A&A* **449**, 369–378 (2006).

Solar Prominence Models - A New Perspective via the Theory of Dynamical Systems

Vasile Mioc, Cristiana Dumitrache

Astronomical Institute of the Romanian Academy, Bucharest, Romania, e-mail: vmioc@aira.astro.ro; crisd@aira.astro.ro

Abstract. We tackle the 2D models of Kuperus-Raadu and Kippenhahn-Schluter, extended by Malherbes and Priest, via the geometric methods of the theory of dynamical systems. We treat the models as phase portraits, and discuss them from the standpoint of mathematics only, without regard to physical significance. Some phase portraits are very intricated and we raise questions about them. We also suggest possible phase portraits. This paper tries to draw attention to new tools, used in celestial mechanics, as useful for the solar physics too and for a suitable interdisciplinary cooperation.

Keywords: solar physics, theory of dynamical systems, prominences,current sheets.
PACS: 96.60.Na, 96.60.P-

INTRODUCTION

Solar prominence models as current sheets, which win against gravity via potential magnetic fields, were proposed half a century ago [1]. Since then, many other models (especially 2D) were proposed, and treated both analytically or numerically. Priest [4] extended the classical models KS [1] and KR [2], and divided them in two important classes, NP (normal polarity) and IP (inverse polarity), which represent two known basic geometries supporting prominences against gravity.

The goal of this paper is to present other methods of investigation of the solar prominence magnetic fields, methods that are complementary to the previous ones. We start from the paper by [3] (hereafter MP). Our basic idea is to extract more information from the models, and to interpret the information in terms of motion. So, we shall interpret the models tackled by MP in terms of *phase portraits*, discuss them only from the mathematical standpoint, and formulate some suggestions and possible developments.

We approach the NP-type and IP-type models discussed by MP. Even these simple magnetostatic analytical solutions offer rich portraits of motions, which we translate in terms of phase portraits and treat from the viewpoint of the theory of dynamical systems.

In section 2 we emphasize the basic equations from which MP started. These equations were modified by these authors in order to model current sheets recoverable via observations.

Section 3 treats the IP-type models. There are more or less intricate phase portraits, but problems to be studied deeper are emphasized: nature of the saddle at the base of the current sheet, bifurcations created by this saddle (Fig. 3), or the existence of a homoclinic orbit generated by the top of the current sheet, which could entail quasiperiodic and periodic orbits (magnetic field lines) and also a hidden center.

CP934, *Flows, Boundaries, Interactions*
edited by C. Dumitrache, V. Mioc, and N. A. Popescu

In section 4 we investigate the phase portraits corresponding to the NP-type models. There also are more or less complicated phase portraits and the same problems appear: the existence, or not, of a homoclinic orbit generated by the top point of the prominence current sheet, entailing the existence of quasiperiodic and periodic orbits inside and a hidden center (Fig. 6). Starting from Fig. 6, we suggest the possibility of another choice of the parameters, able to lead to much richer phase portraits (Fig. 7). In these portraits, the existence of quasiperiodic and periodic orbits is ensured even without a center.

Section 5 resumes the most important questions and formulates some conclusions and suggestions.

We hope that such rather unusual methods could enrich the tools of the solar physics, even by suggesting new phase portraits, which could model observed situations and structures.

BASIC EQUATIONS

To model magnetic fields able to support 2D current-sheet prominences, MP resorted to holomorphic functions of the complex variable $\xi = x + iy$. They defined the complex magnetic potential field in the (x, y)-plane as

$$B(\xi) = B_y + iB_x, \tag{1}$$

where the horizontal x-axis is normal to the prominence sheet, while the vertical y-axis is in the opposite direction to gravity.

As departure point for their investigations, MP used two simple functions: one for an infinite current sheet in the y-direction,

$$B_A(\xi) = B_0\sqrt{p^2 + \xi^2}/\xi, \tag{2}$$

the other for a finite current sheet in the same direction,

$$B_B(\xi) = B_0\sqrt{(p^2 + \xi^2)(q^2 + \xi^2)}/\xi. \tag{3}$$

Here $(B_0, p, q) \in \mathbb{R}^3_+$; p and q stand for the heights of the base and top of the current sheet from the x-axis, respectively. Starting from 2 and 3, MP constructed more intricate potential fields, intended to recover magnetostatic solutions of IP-type and NP-type of prominences.The expressions of these fields also include the parameters $(B_1, h) \in \mathbb{R}^2_+$ (see below).

In the next section we tackle the IP-type and NP-type prominence fields, as generalized by MP, from a single perspective: discussion of the associated phase portraits. As pointed out in the introductory section, we adopt the standpoint of a mathematician: we examine the phase-plane curves and comment them using the tools of the theory of dynamical systems, without trying (for the moment) a physical interpretation.

For our strictly mathematical comments, we do not care about the nature of the field or about the solar zones in which the motions take place. We consider only the phase-plane structure. By Cauchy's theorem, for every set of initial conditions, a solution exists and is unique. An important point to be emphasized is the time interval on which a solution

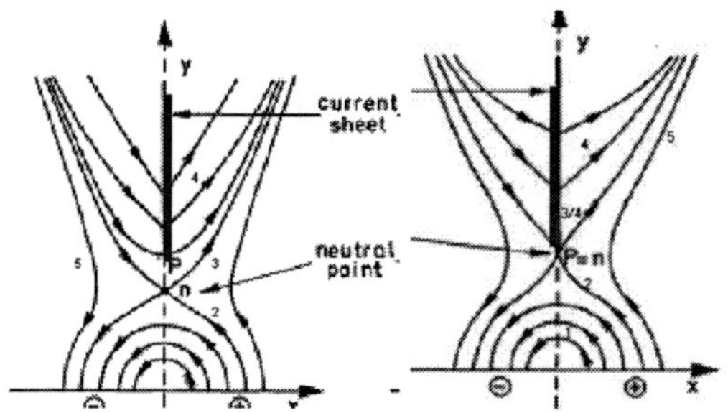

FIGURE 1. Infinite current sheets of IP-type. (a) $h < p$; (b) $h = p$

is defined. Since every solution is related to the lifetime of the current sheet (which is finite in both past and future), it is clear that all solutions encounter singularities (instants when they cease to exist).

Another important point consists of a convention as regards the terminology (just to use the terms of the classical theory of dynamical systems). By abuse, we shall call the solutions that start from/end on the x-axis *ejection/collision orbits*. The solutions that come from/tend to infinity will be called *capture/escape orbits*. A neutral point will be considered as an *equilibrium*. The current sheet will be considered as a *continuum of equilibria* in the phase plane (halfline/segment for infinite/finite sheet).

IP-TYPE MODELS

For an infinite current sheet, MP modified (2) to

$$B_1(\xi) = B_0\sqrt{p^2+\xi^2}/\xi + iB_1(\xi - ih)/\xi, \tag{4}$$

obtaining the phase portraits sketched in Fig. 1 (thick halfline; segment of line for finite sheets, see below).

In panel (a), curves (1) represent closed orbits that eject from collision and then end in collision. Curves (4) and (4') are open orbits that eject from an equilibrium and tend to infinity (4), or conversely (4'). The unstable equilibrium (UE), at $y = n < p$, is a genuine saddle. The trajectories (2) and (2') are, respectively, orbits that eject from collision and tend asymptotically to UE (2), or conversely (2'). The open curves (3) and (3') eject asymptotically from UE and tend to infinity (3), or conversely (3'). Lastly, curves (5) and (5') eject from collision and tend to infinity (5), or conversely (5'). Some "primed" curves are not numbered on the figure, but they can be immediately identified.

In panel (b), the notation for the phase curves has the same significance as in panel (a). Nevertheless, there is a difference: in panel (a), the saddle UE and the lowest degenerate equilibrium are different ($n < p$), while in panel (b) they are identified. That is why we used the notation (3/4), (3'/4') in panel (b). In panel (a), every equilibrium on the current sheet has one stable manifold and one unstable manifold. Only the saddle UE has (obviously) two stable manifolds and two unstable manifolds. It is the same as regards

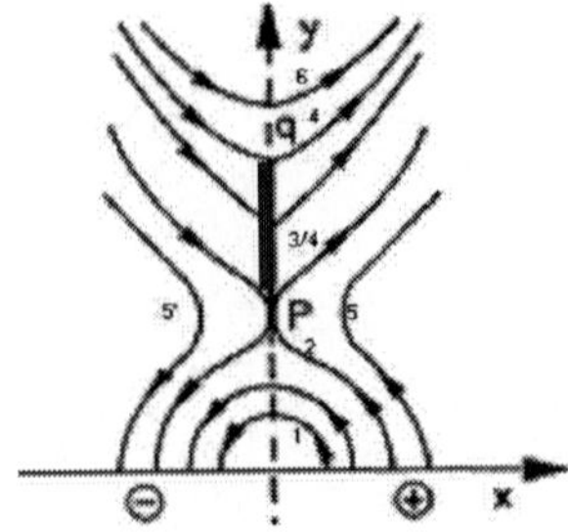

FIGURE 2. Finite current sheets of IP-type modeled by (5)

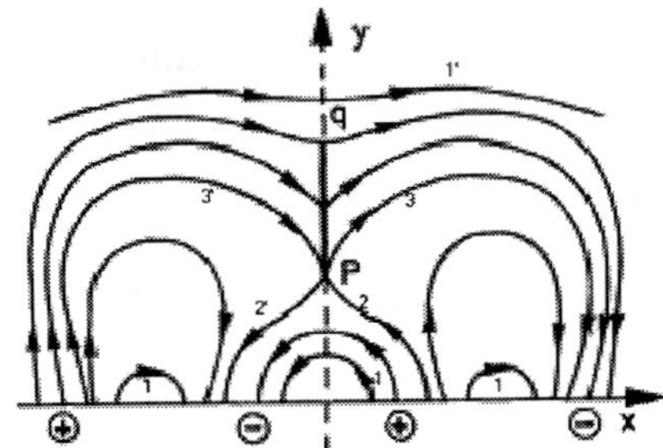

FIGURE 3. Finite current sheets of IP-type modeled by (6)

panel (b), except the point $p = n$, which also benefits of two stable manifolds and two unstable manifolds. We consider that this point has to be studied closer.

For a finite current sheet, MP considered three models: with open field lines towards the corona, with closed field lines down to the photosphere and with helical structure.

The first model uses a change of (2) as follows:

$$B_2(\xi) = B_0\sqrt{(p^2+\xi^2)(q^2+\xi^2)}/\xi + B_1(\xi - ip), \tag{5}$$

and provides the phase portrait given in Fig. 2.

The description and interpretation of the phase portrait are analogous to those of Fig. 1b. There are, however, some differences. There is no more a halfline of equilibria, but only a segment of length $q-p$. In addition, there appear the trajectories (6), which come from infinity and then tend back to infinity. The properties of the point p remain to be further investigated.

The second model uses a change of (3) as follows:

$$B_3(\xi) = -B_0\sqrt{(p^2+\xi^2)(q^2+\xi^2)}/[\xi(\xi+ih)^2] - B_1(\xi - ip)/[\xi(\xi+ih)]. \tag{6}$$

It yields the phase portrait given in Fig. 3 (provided $q \leq 2p$).

The phase portrait is more complicated. There are the already known orbits 1 (ejection-collision), below the point q, but coexisting with the same type of orbits (1')

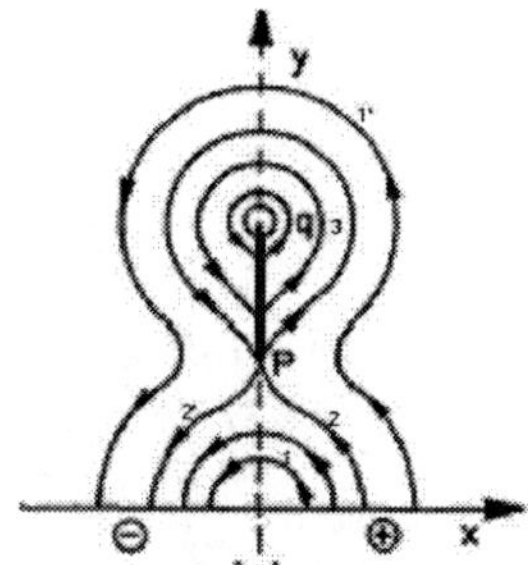

FIGURE 4. Finite current sheets of IP-type modeled by (7)

that go beyond the point q. Orbits (2) and (2') have the same significance as above. Curves (3) and (3') start from an equilibrium and end in collision (3), or eject from collision and tend to an equilibrium (3'). Remark that curves (3') and (2) have the same significance as regards trend; it is the same for the curves (3) and (2').

The point p still has the look of a saddle, but *an unusual one*. All orbits that come from/end in it have the other endpoint in collision/ejection. This creates the unusual evolution of the curves (1) that do not intersect the x-axis. The separation (bifurcation) is clearly produced by the curves (2) and (2') on each halfplane $x > 0$ and $x < 0$.

The third model seems the most interesting for us. Formula (3) was turned by MP to

$$B_4(\xi) = -B_0\sqrt{(p^2+\xi^2)(q^2+\xi^2)}/[\xi(\xi+ih)^2] - B_1(\xi - ip)/[\xi(\xi - iq)]. \quad (7)$$

The corresponding phase portrait is given in Fig. 4, with $q/p < B_1/B_0$.

Of course, there exist already known trends as those illustrated by trajectories (1) and (1'), (2) and (2') (see the second IP-type model). But the phase portrait is much more interesting now. The point p still looks like a saddle, but it now generates a homoclinic orbit (3). Moreover, an infinity of equilibria on the current-sheet segment generate homoclinic orbits.

One problem to be digged deeper is the appearance of chaos (better said, unpredictability), given the existence of homoclinic orbits.

Another problem is the existence of quasiperiodic and periodic orbits. It is not clear to us whether the orbits in the more and more small neighborhood of q tend to the point $y = q$ (filling densely the interior of the homoclinic orbit generated by p) or still allow a homoclinic orbit originating in and tending to this point. If such a homoclinic orbit exists, there are infinitely many quasiperiodic and periodic orbits inside it. All such trajectories surround a center (stable equilibrium).

NP-TYPE MODELS

The NP-type models tackled by MP imply three variants, too: with infinite current sheets in the y-direction, with finite currents sheets in same direction, and with helical structure. The first two models were concretized by MP in the formulae:

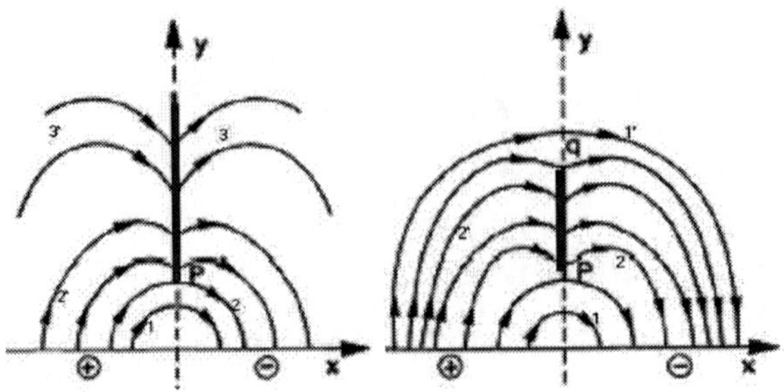

FIGURE 5. Current sheets of NP-type. (a) infinite; (b) finite

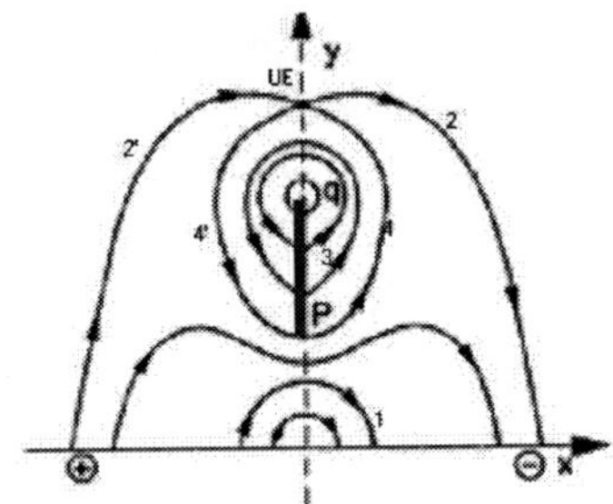

FIGURE 6. Finite current sheet of NP-type modeled by (10)

$$B_5(\xi) = iB_0\sqrt{p^2+\xi^2}/[\xi(\xi+ih)] - B_1/\xi; \tag{8}$$

$$B_6(\xi) = -B_0\sqrt{(p^2+\xi^2)(q^2+\xi^2)}/\,[\xi(\xi+ih)^2] - B_1/\xi, \tag{9}$$

respectively (see Fig.5).

The interpretation of the phase curves plotted in Fig. 5 is already known (compare with the above commented phase trajectories in the NP-case). There exist: a continuum of equilibria, ejection-collision curves (of two kinds), capture-escape curves, as well as orbits that eject from collision and tend to equilibria, or conversely.

Finally, the last modification of formula (3) proposed by MP reads

$$B_7(\xi) = -B_0\sqrt{(p^2+\xi^2)(q^2+\xi^2)}/[\xi(\xi+ih)^2] - (B_1/\xi)\,[(\xi+ih')/(\xi-iq)] \tag{10}$$

with $q^2p/(h^2h') < B_1/B_0 < 1$ and $h'q > h^2$.

The corresponding phase portrait looks as in Fig. 6.

Here we have well-known orbits, as 1. We still have orbits that come from/end to an equilibrium (ejecting or ending from collision; e.g.(2), (2')). Examples of homoclinic orbits are given by the curves 3. A new feature of this phase portrait is the existence of

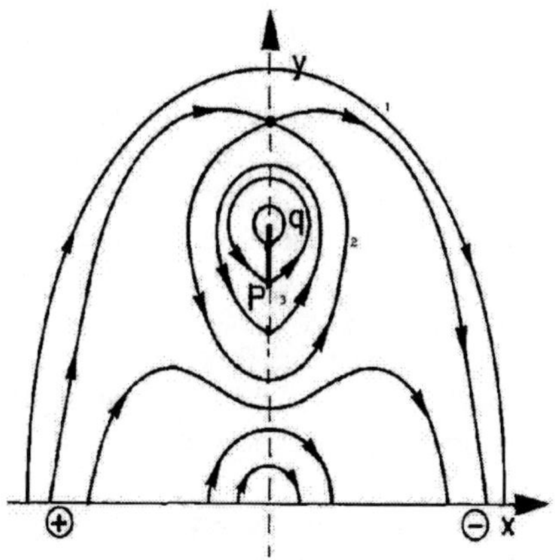

FIGURE 7. Is such a NP-type phase portrait possible?

heteroclinic orbits, which come from p to UE (4), or conversely (4'); both trends are asymptotic.

We have to formulate two questions within this framework. First: is there a homoclinic orbit emerging from and ending in q (see the final of section (3))? This would imply the existence of quasiperiodic and periodic orbits inside this loop, which cannot exist inside the other homoclinic loops between q and p. Second: could we find a combination of parameters to imply the existence of an "outer" ejection-collision orbit and a genuine homoclinic orbit moving down under p and sheltering quasiperiodic and periodic orbits? We sketch such a phase portrait in Fig. 7.

In this figure we denoted: (1)- "outer" ejection-collision orbit, (2) - genuine homoclinic orbit, (3)- quasiperiodic and periodic orbits. Observe that these last orbits are confined inside an annulus bounded by the homoclinic curve (2) and the homoclinic curve generated by p. Such a situation forbids the existence of a stable equilibrium (center), proper to the quasiperiodic and periodic orbits (3), around which these ones evolve.

CONCLUSIONS

Using the qualitative geometric methods of the theory of dynamical systems, we investigated only the models of solar prominences proposed by MR. We commented the phase portraits without using any physical interpretation. We sketch below some concluding remarks and suggestions.

Some of the phase portraits we examined display a rich variety of phase curves. We mention especially those presented in Figs. 3, 4 and 6.

The properties of the continuum of equilibria that form the current sheet still have to be studied and discussed. We refer especially to the point q in Figs. 4 and 6: is it generating a homoclinic orbit (in its immediate neighborhood) or not? This entails the problem of existence or nonexistence of quasiperiodic and periodic orbits inside the homoclinic orbit and of a corresponding center (stable equilibrium), too.

The properties of the point p as a saddle in Figs. 1b, 2, 3 and 4, must also be investigated a little bit deeper.

The phase curves 2 and 2' in Fig. 3 create bifurcations in both the halfplane $x > 0$ and the halfplane $x < 0$. This situation also deserves to be examined.

In some cases, modeled or only suggested, homoclinic orbits appear. It would be useful to search for chaotic behavior (in Poincaré's sense) in these cases (e.g. via Melnikov's integral). What kind of homoclinic tangle could be produced by the configuration presented in Fig. 4? This is another important question.

As conclusions with general character, we formulate:

Starting only from MP phase portraits, new structures of the phase plane can be proposed. They must be supported by realistic variations of the parameters that feature the involved models.

Our approach is only a geometric one, and needs the support of physical interpretation of the phase curves seen, commented, or proposed.

This requires a closer cooperation between solar physicists and celestial mechanicians, in order to offer new or less usual tools to solar features study.

ACKNOWLEDGMENTS

The authors thank Professors Eric Priest and Jean-Marie Malherbe for the permission of reproducing and using some of their MP-figures. They also thank Professor Terry Forbes for valuable discussions and comments.

REFERENCES

1. Kippenhahn R., Schlüter A.: 1957, Zs. Ap. 43, 36
2. Kuperus M., Raadu M.A.: 1974, Astron. Astrophys. 31, 189
3. Malherbe J.H., Priest E.R.: 1983, Astron.Astrophys. 123, 80
4. Priest E.R. (ed.): 1988, Dynamics and Structure of Quiescent Solar Prominences, Springer-Verlag, New York

The Power of Time-Frequency Representations

Adrian Oncica

Astronomical Institute of the Romanian Academy, Bucharest, Romania, e-mail: aoncica@aira.astro.ro

Abstract. Unlike most scientific disciplines, astronomical model-building must typically rely on observations alone. No direct manipulation of the system or classical physical experiment is possible. Because of this, there has long been a close connection between astronomy and the study of time series. In fact the modern form of latter discipline arguably starts with Yule's analysis in 1927 of sunspot data series and the introduction of what we now call an ARMA model. In astrophysics we deal generally with inverse problems. Observations of some phenomena have to be understood even we have little knowledge on its driving mechanism or where exactly it operates. The problem is then to discover the nature of the underlying physics or geometry and how it produces the observed effects. For many signals in nature the frequency content changes in time as the physical processes producing the signal are time dependent. The tool for describing time-varying spectrum is called time-frequency analysis and saw significant progress in recent years. We intend to give a short review of the idea and its development and to point out qualities and drawbacks. Some examples from both long term and short term solar related data will be given. Finally we will focus on some long term Ulysses data.

Keywords: time series, interplanetary magnetic field, Ulysses mission
PACS: 96.50.Wx, 96.60.vg, 95.75.Wx

THE TIME-FREQUENCY PLANE

Localization and Spreading

We first meet the signal $x(t)$ in the *Shannon representation*, as a time series of samples as provided by the measuring instrument. It tells the values of the signal and when these values occur. Next we can build the *Fourier representation* through: $X(\nu) = \int_{-\infty}^{+\infty} x(t) e^{-j2\pi\nu t} dt$. It tells us what spectral content our signal has but not when its components occur.

Lets see now what if we look at $x(t)$ and $X(\nu)$ as probability distributions. Lets look at their mean and dispersion in the next table.

TABLE 1.

	localization	spreading
time	$t_m = \frac{1}{E_x}\int_{-\infty}^{+\infty} t\|x(t)\|^2 dt$	$(\Delta t)^2 = \frac{4\pi}{E_x}\int_{-\infty}^{+\infty}(t-t_m)^2\|x(t)\|^2 dt$
frequency	$\nu_m = \frac{1}{E_x}\int_{-\infty}^{+\infty} \nu\|X(\nu)\|^2 d\nu$	$(\Delta\nu)^2 = \frac{4\pi}{E_x}\int_{-\infty}^{+\infty}(\nu-\nu_m)^2\|X(\nu)\|^2 d\nu$

We assume here that $E(x)$, the energy of the signal, is bounded:

$$E_x = \int_{-\infty}^{+\infty} |x(t)|^2 dt = \int_{-\infty}^{+\infty} |X(\nu)|^2 d\nu < +\infty \tag{1}$$

CP934, *Flows, Boundaries, Interactions*
edited by C. Dumitrache, V. Mioc, and N. A. Popescu

In this way a signal can be characterized in the time-frequency plane by its mean position (t_m, ν_m) and a domain of main energy localization $(\Delta t, \Delta \nu)$ whose area is proportional to the time-bandwidth product $\Delta t \times \Delta \nu$. One interesting property of this product is that it is lower bounded $(\Delta t \times \Delta \nu \geq 1)$. This constraint, known as the *Heisenberg-Gabor inequality*, illustrates the fact that a signal can not have simultaneously an arbitrarily small support in time *and* in frequency.

Atomic Decomposition

As the Fourier transform is not suited for multi-component and/or non-stationary signals, especially noisy ones, one has to consider bidimensional (in time and frequency) functions. The simplest illustration of this concept is the *short-time Fourier transform* (STFT):

$$F_x(t,\nu;h) = \int_{-\infty}^{+\infty} x(u)h^*(u-t)e^{-j2\pi\nu u}du \tag{2}$$

where $h(t)$ is the short-time analysis window.

So the time-frequency plane is covered with "atoms" which have the time resolution of $h(t)$ and the frequency resolution of $H(\nu)$. Since they are reciprocal through the Fourier transform, the Heisenberg-Gabor inequality follows. If the window is gaussian (if in time then in frequency also, a nice property of the Fourier transform) we are in the *Gabor representation* [3].

We end this inventory with the *continuous wavelet transform* (CWT):

$$T_x(t,a;\Psi) = \int_{-\infty}^{+\infty} x(s)\frac{1}{\sqrt{|a|}}\Psi^*(\frac{s-t}{a})ds \tag{3}$$

where $\Psi(s)$ is the, zero mean, mother wavelet, generator (through time t translation and scale a dilatation) of the family. In contrast to the STFT, which uses a single analysis window, the CWT uses short windows at high frequencies and long windows at low frequencies (allowing to see the forest *and* the trees).

Energy Decomposition

If we consider the square modulus of STFT (from eq.2) we obtain the spectral energy density $S_x(t,\nu)$ or *the spectrogram*. Assuming the window h of unit energy, the spectrogram satisfies:

$$\int_{-\infty}^{+\infty}\int_{-\infty}^{+\infty} S_x(t,\nu)dtd\nu = E_x \tag{4}$$

so we can see the spectrogram as a measure of the energy distribution in the time-frequency domain (see eq. 1) also. Deriving from STFT it has the same poor resolution properties, namely the trade-off between time and frequency resolution. Being quadratic it gives rise to interference terms if more than one signal is present (this includes noise).

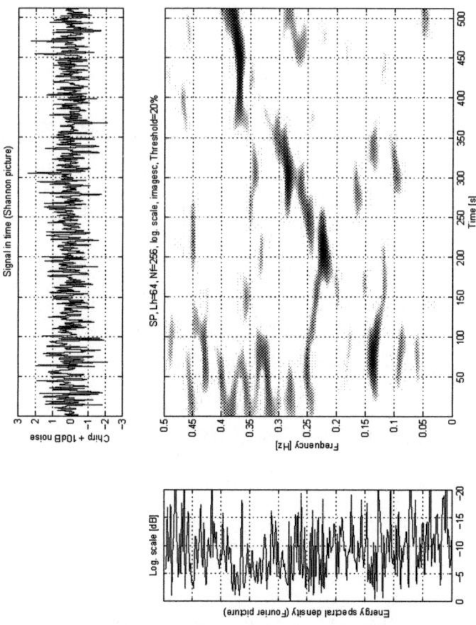

FIGURE 1. Chirp signal embedded in +10dBW gaussian noise

Even so its ability to extract signal from noise is obvious in the next figure. It represents a chirp signal (linear frequency modulated) embedded in a +10dBW gaussian noise.

There is no indication of the signal in the Shannon representation above. It seems to be pure noise. There is no spectral feature in the Fourier representation at left also. Nevertheless, the joint time-frequency representation reveals with no doubt the presence of the chirp signal. The reason is that the power of the gaussian noise is distributed across the entire time-frequency plane. In the meantime the power of the signal remains concentrate in its narrow strip.

The following step is to look for a joint (time and frequency) distribution $\rho_x(t,\nu)$ (replacing S_x in eq. 4) and satisfying two marginal properties (see eq. 1 also):

$$\int_{-\infty}^{+\infty} \rho_x(t,\nu)dt = |X(\nu)|^2 \tag{5}$$

$$\int_{-\infty}^{+\infty} \rho_x(t,\nu)d\nu = |x(t)|^2 \tag{6}$$

As ρ_x is not unique we can impose a number of desirable mathematical properties, covariance being fundamental.

A time-frequency energy distribution with lots of desirable properties is *the Wigner-Ville distribution* (WVD) defined as:

$$W_x(t,\nu) = \int_{-\infty}^{+\infty} x(t-\tau/2)x^*(t+\tau/2)e^{-j2\pi\nu\tau}d\tau \tag{7}$$

or:

$$W_x(t,\nu) = \int_{-\infty}^{+\infty} X(\nu - \xi/2)X^*(\nu + \xi/2)e^{-j2\pi\xi t}d\xi \tag{8}$$

This distribution satisfies the energy conservation (eq. 2), and marginal conditions (eqs. 5, and 6). Is real valued and covariant to translation and dilatation both in time and in frequency. It is compatible with filtering and modulation. It has many others useful properties but as quadratic it is prone to interference terms. Throw in eq. 7 a window $h(t)$ and we get *the pseudo-Wigner-Ville distribution* (PWV). Some properties are lost but time interferences diminish. A similar procedure can be applied in the frequency domain to obtain the *smoothed-pseudo-Wigner-Ville distribution* (SPWV) with similar results regarding the frequency interferences. Both procedures improve the behavior regarding interferences at the expense of loosing resolution (both in time and frequency). The last step is the reassignment procedure which recovers part of the lost resolution. For further information see [2].

THE REAL TIME DATA

Collecting the Data

The problem is to get a hint of the nature of the underlying physics or geometry and how it produces the observed effects from the spectrum and its behavior. "Spectrum is geometry". "Listen" to the data or look at its spectrum and if you "hear": 1, 2, ... you can think in terms of linear harmonics. Or if you "hear": 1.0, 2.4, ... you may think cylindrical.

MHD processes (base of some solar radio bursts, especially type IV) contain signatures of specific emission mechanisms (gyromagnetic, plasma interactions, Langmuir waves, loss-cone, and others). All of them have specific behavior both in time and frequency through parameters such as: bandwidth (large or narrow), variability (fast, slow, or none); amplitude (constant, decaying, variable) [1] [4] .

Most of long term solar observations may contain the spectral signature of the solar rotation. If it is present is the connection clear? Or if it is missing why? During no more than the elaboration of this work, an abnormal behavior of the spectrum pointed to some errors in the original and official Ulysses data.

Sometime you can meet frequencies which should not be present (poor calibration, erroneous data, contamination, and others). For example a one year period in the solar radio data is certainly due to poor calibration with elevation angle [5].

We choose to inspect some Ulysses data (magnetic field and plasma parameters) during the entire mission up until near present. In the next figure we gathered some of the data we choose for our next analysis. In the first two rows are the heliographic distance (in AU) and the heliographic latitude (in degrees) to have an idea where the spacecraft is. In the next two rows we put the the two solar activity tracers: the *International Sunspot Number* (R_i) and the *10 cm Solar Radio Flux* (F_{10}).

We shall concentrate at the beginning onto the ecliptic crossing. The first took place in 1995 during the decaying end of the Solar Cycle 22. The second took place in 2001 during the maximum phase of the Solar Cycle 23. In the magnetic data plots they are

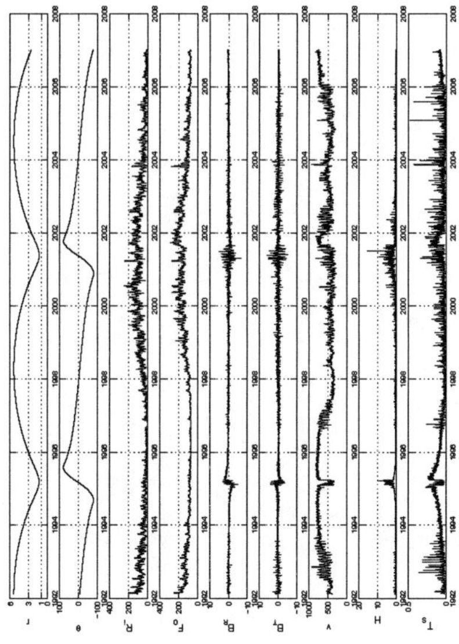

FIGURE 2. The 15 years of Ulysses data and more

both evident even though quite different. On the other hand the plasma parameters (v, ρ_H, and T_S) enjoy a clear and expected behavior during the first crossing (in 1995, at almost solar minimum) but a nearly chaotic behavior during the second crossing (in 2001, at solar maximum).

Next we shall look at the strong oscillations which catch the eye in the solar wind speed plot during 1992-1993. Are the ones in T_S of the same nature? Are they present in the other plots? What could be their interpretation as long as the spacecraft is some 5 AU away? We continue with the same kind of fluctuations during 1996-1997. Ulysses is again at about 5 AU but this time a few degrees above the ecliptic.

As for what spectral patterns we look for we find them in solar activity behavior. In the R_i dynamic spectrum we ca see the evident signature of the solar rotation (patches at 1 cycle /Bartel rotation). In what follows we will call this frequency simply F_B.

Looking after Patterns

First lets look at the magnetic field variability. In 1995, when the ecliptic is crossed near the solar minimum, both B_R and B_T display no trace of F_B but the first harmonic is present. In 2001, when the ecliptic is crossed at solar maximum, F_B is quite strong and the first harmonic also present.

In the plasma parameters data F_B is also present but less distinct. It may be of no surprise that it is it who is so manifest in the solar wind speed. What may be curios is its presence even during the 1995 solar minimum. What may be interesting to investigate further is why F_B *and its first harmonic* is so evident in the data collected when Ulysses

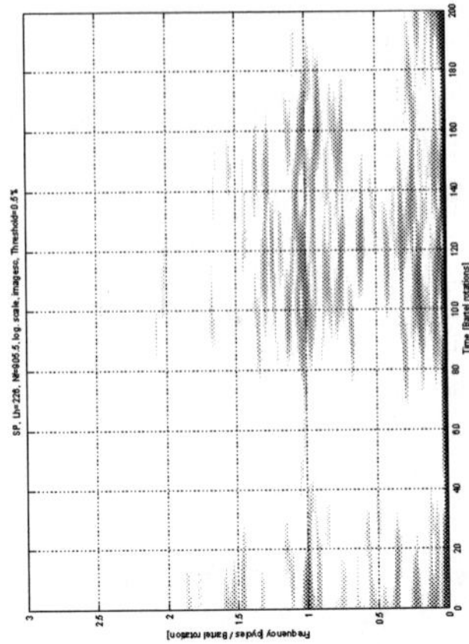

FIGURE 3. The time-frequency spectrum of R_i

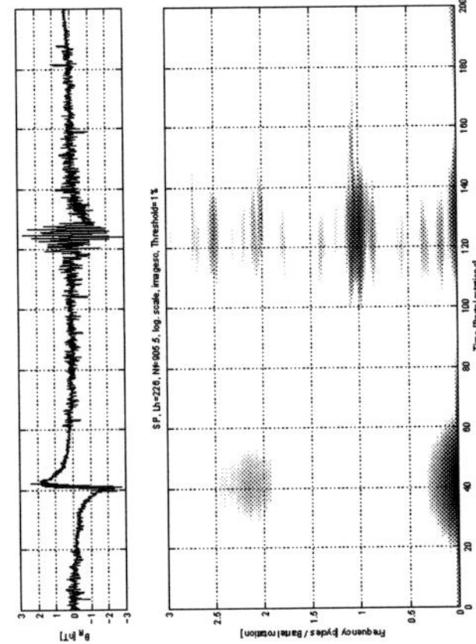

FIGURE 4. The radial component of the magnetic field B_R

is near Jupiter orbit.

CONCLUSIONS

At this point it is difficult to draw conclusions. First the presence of F_B at ecliptic crossing (at some 1 AU) must be of no surprise. A next step can be a quantitative analysis of this presence.

When the same oscillations appear in the data taken at some 5 AU may be not so easy to explain. Moreover we used *time* series for inspection but our detectors are traveling through space so they map also spatial variability. Can this variability have a *spatial* periodicity of some few heliographic latitude degrees instead of F_B?

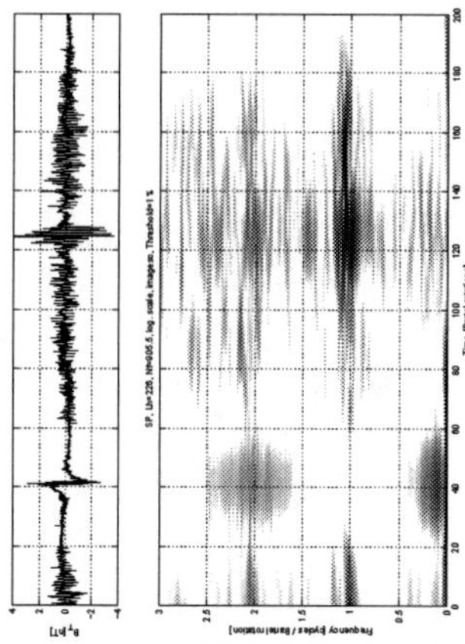

FIGURE 5. The transversal component of the magnetic field B_T

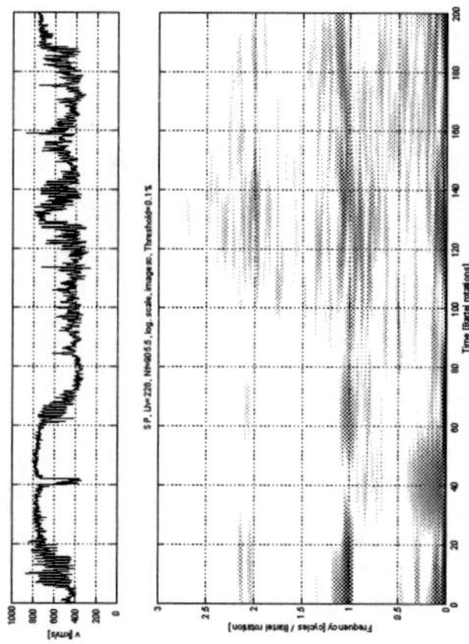

FIGURE 6. The solar wind plasma speed

The answers for this question and some who may rise should be looked in a more detailed investigation of spectral components and will be the task of future work.

ACKNOWLEDGMENTS

The author wish to thank Francois Auger Patrick Flandrin Paulo Goncalves and Olivier Lemoine from CNRS France for their "Time Frequency Toolbox" and to all those people an teams who made the Ulysses Mission data freely available.

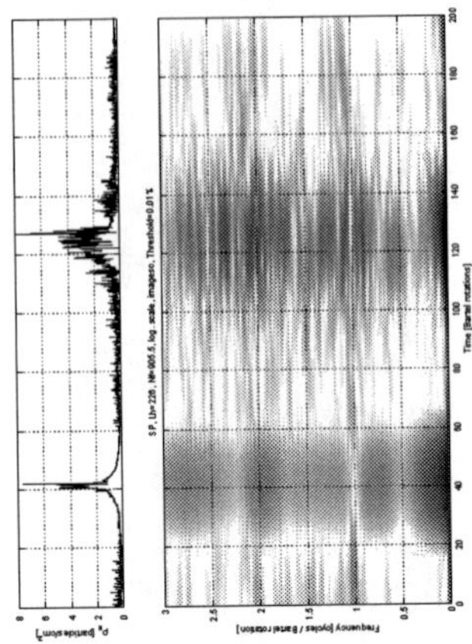

FIGURE 7. The proton density

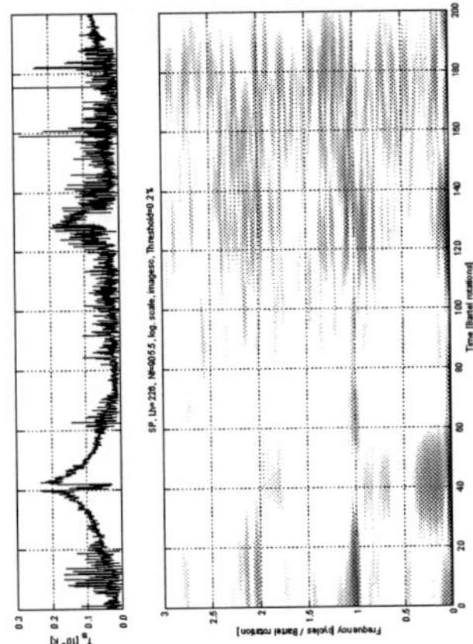

FIGURE 8. The plasma temperature T_S

REFERENCES

1. M. J. Anschwanden, *SolPhys* **111**, 113-136 (1987)
2. F. Auger, P. Flandrin, P. Goncalves, O. Lemoine, "Time Frequency Toolbox" CNRS (1996)
3. D. Gabor, *J.IEE* **93**, 429-434 (1946)
4. A. Oncica, *Rom, Astron.J*, **8**, 24-29 (1998)
5. A. Oncica, *Rom. Astron.J*, **10**, 62-69 (2000)

Numerical simulations of the initiation and the IP evolution of coronal mass ejections

C. Jacobs, S. Poedts, B. van der Holst, G. Dubey and R. Keppens

CPA/K.U.Leuven, Celestijnenlaan 200 B, 3001 Leuven, Belgium

Abstract.
We present recent results from numerical simulations of the initiation and interplanetary (IP) evolution of Coronal Mass Ejections (CMEs) in the framework of ideal magnetohydrodynamics (MHD). As a first step, the magnetic field in the lower corona and the background solar wind are reconstructed. Both simple, axisymmetric (2.5D) solar wind models for the quiet sun as more complicated 3D solar wind models taking into account the actual coronal field through magnetogram data are reconstructed. In a second step, fast CME events are mimicked by superposing high-density plasma blobs on the background wind and launching them in a given direction at a certain speed. In this way, the evolution of the CME can be modeled and its effects on the coronal field and background solar wind studied. In addition, more realistic CME onset models have been developed to investigate the possible role of magnetic foot point shearing and magnetic flux emergence/disappearence as triggering mechanisms of the instability. Parameter studies of such onset models reveal the importance of the background wind model that is used and of the initiation parameters, such as the amount and the rate of the magnetic flux emergence or the region and the amount of foot point shearing.

Keywords: CME initiation, CME evolution, numerical simulation, 3D ideal MHD
PACS: 96.50.Ci, 96.50.Uv, 96.50.Tf, 96.60.ph, 96.60.qf, 96.60.-j, 96.60.Hv

MOTIVATION

Coronal Mass Ejections (CMEs) belong to the most violent and fascinating events in the solar system. These events involve large-scale changes in the coronal structure and significant disturbances in the solar wind. Especially the massive, fast CMEs are interesting to study as these events cause shocks that propagate through the interplanetary (IP) space. In these IP shock waves energetic particles continuously accelerate giving rise to gradual solar energetic particle events (SEPs). As a result, CMEs and the CME generated shock waves play a key role in the so-called space weather. Mathematical modelling of these dramatic solar eruptions might help us to get a better insight in the evolution and impact of CMEs and is essential for a better prediction and understanding of the physical processes involved in the space weather. Due to the different time and length scales, numerical modelling of solar transients is a challenging task and asks for a lot of computational power and the continuous improvement of numerical techniques and physical models. During the last couple of years significant progress has been made in the field of computational magneto-fluid-dynamics, applied to solar related phenomena. This progress is due to the improved computer facilities and also to the inclusion of more realistic, data-driven boundary conditions in the models.

In this paper the recent progress made at the Centre for Plasma-Astrophysics (K.U.Leuven) in numerically simulating the initiation and IP evolution of CMEs is

CP934, *Flows, Boundaries, Interactions*
edited by C. Dumitrache, V. Mioc, and N. A. Popescu

presented. All results are obtained in the framework of ideal magnetohydrodynamics (MHD). The ideal MHD equations are solved using a finite volume code derived from the Versatile Advection Code (VAC)[1]. This code supports parallel computations on distributed memory machines, using the Message Passing Interface (MPI). In the results presented here the magnetic field is kept divergence-free by using the vector potential on the nodal points in stead of working with the cell-centred magnetic field components, as suggested by Balsara and Spicer [2]. The equations are solved in a spherical geometry [(r,θ,φ)-coordinates] where the computational domain is restricted to $1-30R_\odot$ in the radial direction and covers the region between north and south pole of the Sun, i.e. $0 \leq \theta \leq \pi$. For the 3D simulations a full sphere is modeled where the cells are equidistant in the radial direction ($0 \leq \varphi \leq 2\pi$). In the next section the current models for the solar background wind are discussed. Both axisymmetric (as well in 2.5D as 3D configuration) models, representing the solar wind at solar minimum conditions, as more complicated 3D corona models are considered. Next, we briefly discuss our current CME initiation models and present recent results on 3D IP CME evolution. Finally, some preliminary conclusions are drawn.

SOLAR WIND MODELS

In this section our different types of solar wind models are discussed. A first series of simple models is based on the assumption of an ignorable azimuthal direction and a dipole magnetic field structure. These models are an approximation for the solar wind under solar minimum conditions. Next, a model including a more complicated magnetic field structure is presented, using input from magnetograms as a boundary condition.

2.5D solar wind models

Our axisymmetric models are reconstructions of solar wind models often used in the literature. A first model is the classical polytropic wind model, where a relation ρp^γ is assumed, with ρ the density, p the pressure and γ the polytropic index. In this way the energy equation can be omitted, simplifying the equations. For the polytropic index a value of $\gamma = 1.05$ was chosen. This model is fairly popular in the literature in spite of its clear shortcomings, e.g. it is not possible to obtain fast solar wind speeds and the ratio of the densities in the closed and open field line regions is too small. Wind model 2 is similar to the model first developed by Groth et al. [3] and described in more detail by Manchester et al. [4]. In this model, the full set of ideal MHD equations is solved and an extra term of the form

$$Q = \rho q_0 \left(T_0 - \gamma \frac{p}{\rho} \right) e^{-\frac{(r-R_\odot)^2}{\sigma^2}},$$

is added to the energy equation to mimic the effect of radiative losses, thermal conduction and other heating/cooling mechanisms. Here, the ratio of specific heats is $5/3$. The presence of the heating/cooling term causes sharp gradients in the density and velocity profiles (see Fig. 1). This wind model possesses both a fast and a slow wind component, and is in that sense an improvement compared to the polytropic model. The third axisymmetric wind model is again a polytropic model, but now including the effect of Alfvén waves [5], which gives rise to an extra (wave) pressure gradient p_w in the momentum equation, which is determined from

$$\frac{\partial \epsilon_\pm}{\partial t} + \nabla \cdot [(\mathbf{v} \pm \mathbf{v}_A)\epsilon_\pm] = -\frac{1}{2}\epsilon_\pm \nabla \cdot \mathbf{v},$$

where $\epsilon_\pm$ is the average Alfvén wave density for the parallel and anti-parallel propagating waves and $\mathbf{v}_A$ is the Alfvén velocity. The extra Alfvén wave pressure is then given by:

$$p_w = \frac{\epsilon_+ + \epsilon_-}{2}.$$

This wind model has the correct pole-to-equator velocity ratio and has, compared to model 2, a smoother transition going from the equator towards the pole. Figure 1 shows the radial velocity and particle density along the equator and at a constant radius of $30\,R_\odot$ for the three wind models. All three wind models possess a dipolar magnetic field structure and show up-down symmetry over the equatorial plane.

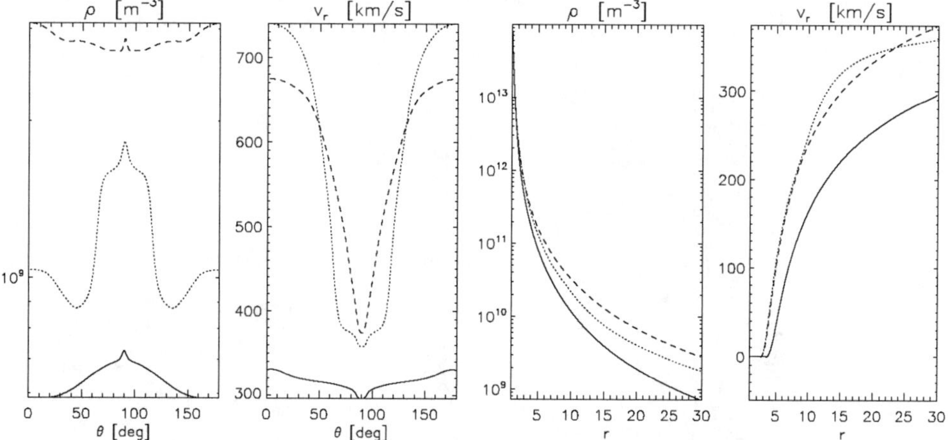

FIGURE 1. The profiles for the density and the radial velocity at $30R_\odot$ (left) and along the equator (right). Solid line: wind model 1, dotted line: wind model 2, dashed line: wind model 3.

These different models where all constructed with the same code, i.e. the same numerical scheme, on the same mesh and with the same initial and boundary conditions, such that they only differ in the physics included in the equations. These models are used for parameters studies to investigate the effect of different solar wind conditions on the initiation and the propagation of CMEs [6, 7].

Full 3D solar wind model

Even at solar minimum the global magnetic field of the Sun is not exact a dipole, and variation in the azimuthal direction is present. A more realistic model for the coronal magnetic field is obtained by including information from magnetograms in the model. In the example shown below, the PFSS reconstruction of Schrijver and DeRosa [8] was used as an initial condition for the magnetic field. Since the magnetic field above sunspots can be rather high, high spatial resolutions are needed to solve the gradients in the magnetic field. However, the small scales and the strong magnetic fields put serious limitations on the numerical time step taken by the code. For this reason, the lower boundary was taken at $1.03Rs$ and the order of the spherical harmonics in the potential field reconstruction was limited to $l = 3, m = 3$. Since this is a preliminary result, again a polytropic relation between density and pressure is assumed. In order to have optimal resolution where needed, we opted for an adaptive mesh refinement (AMR) technique. The AMR-VAC code [9] has recently been improved with an MPI implementation, which will enable us to increase the resolution considerably and to include more spherical harmonics. In Fig. 2 the simulated coronal magnetic field structure at March 29, 2006 is shown. Three levels of refinement were used to obtain this result. The first level had a resolution of $480 \times 32 \times 32$ cells, giving an effective resolution at the third level of $1920 \times 128 \times 128$ cells. The other plot in Fig. 2 shows a longitude vs. latitude map of the radial velocity at a distance of $2.5\,R_\odot$, i.e. the position at which initially the source surface was located, with the location of the current sheet being clearly visible.

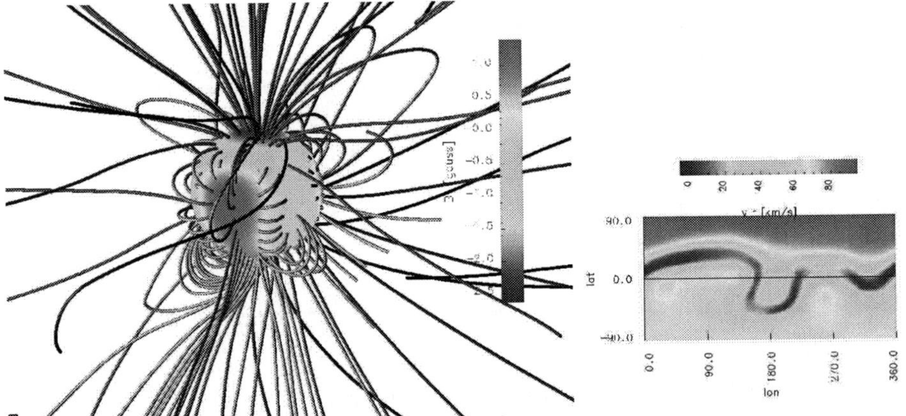

FIGURE 2. Coronal field reconstruction for March 29, 2006. Left: the magnetic field structure. Right: longitude-latitude map of the radial velocity at $2.5\,R_\odot$.

CME INITIATION MODELS

Flux emergence / cancelation models

Jacobs et al. [10] triggered CMEs from the axisymmetric wind models discussed above by shearing the foot points of the magnetic field lines with an extra longitudinal flow profile on the solar surface. The temporal evolution of the magnetic energy, the velocity of the flux rope, and the magnetic helicity shows a dependence on the maximum shear velocity as well as on the background wind model. CMEs can also be triggered by the emergence of additional magnetic flux of the same or the opposite polarity as the overlying magnetic field. Dubey et al. [11] considered a flux rope in a dipole magnetic field that is initially kept stable by means of a line current and studied the dependence of the evolution parameters, such as the velocity and the acceleration, on the flux emergence rate and the total amount of flux. More recent results take into account the background solar wind (wind model 2 in described above) which yields faster CMEs (see Fig. 3).

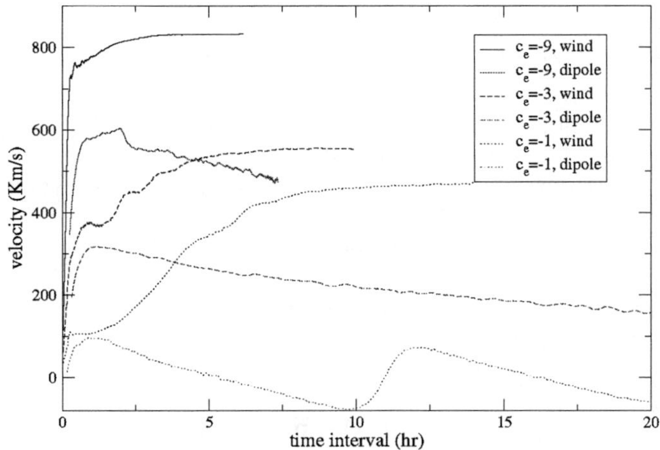

FIGURE 3. Comparison of the velocity evolution for dipole background field and background MHD wind for varying flux emergence rate, i.e. $2\pi c_e \psi_0/\Delta t$ from -1.10×10^{20} to -1.22×10^{19} Mx/s.

'Break out'-Model superposed on solar wind (2.5D)

Another recent and promising result was obtained by considering the 'break out' model superposed on the second wind model described earlier, again triggered by adding extra azimuthal velocity v_ϕ at the solar surface to shear the foot points of the magnetic field [cf. 12]. This is illustrated in Fig. 4. Extensive parameters studies are in progress.

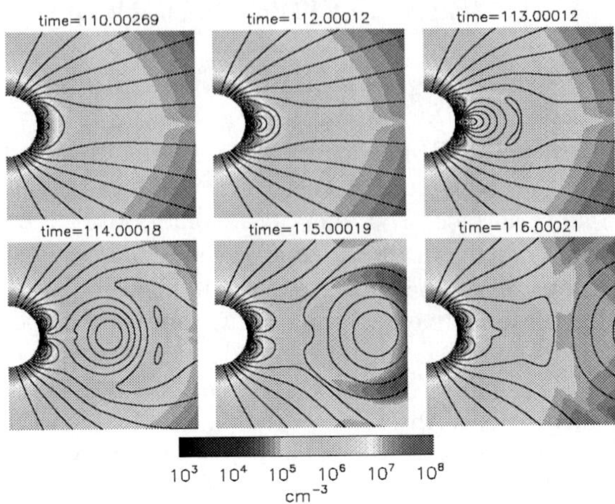

FIGURE 4. Different snapshots of the 'break out' CME including a background wind. The colour indicates density and the black lines represent the magnetic field lines.

INTERPLANETARY EVOLUTION OF CMES

The previous models all assumed axisymmetry. While for the solar corona this might still be a good approach under solar minimum conditions, for CMEs this is definitely no longer a valid assumption. In the next subsection, results of a 3D simulation are compared with similar 2.5D simulations in order to point out the differences and to check the reliability of the 2.5D models. Since we are primarily interested in the CME propagation, the CME will be mimicked by superposing a high-density plasma blob to the background wind. For the background wind model the 3D version of the second model described above is used.

2.5D vs 3D CME propagation

The shocks are generated by launching a high-density plasma blob on the wind at a certain speed $v_{\rm cme}$ in a prescribed radial direction. The velocity and density profile in the initial disturbance both take the form:

$$f = \frac{f_{\rm cme}}{2}\left(1 - \cos\pi\frac{d_{\rm cme} - d}{d_{\rm cme}}\right),$$

where f indicates the added density ρ or radial velocity v_r, $f_{\rm cme}$ is the maximum density $\rho_{\rm cme}$ or radial velocity $v_{\rm cme}$ in the plasma bubble, $d_{\rm cme}$ is the radius of the bubble, and d is the distance to the centre of the blob. For the results presented here, the blob was launched in the equatorial plane. After the initiation, the CME was followed during approximately 7 hrs. To reach this time, about 50 000 iterations were needed in the fully three dimensional computation, giving a total simulation time of about 47 hours on 128 CPUs of the VIC-cluster at the K.U.Leuven. In comparison: the axisymmetric computations needed only around 13 000 iterations (because of the larger time step) and were finished in less than 10 minutes when 32 CPUs were used.

In order to create fast shocks the maximum radial velocity in the blob $v_{\rm cme}$ is about $1000\,{\rm km\,s^{-1}}$ for every simulation. In the case of the 3D simulation the total mass in the blob amounts $\approx 10^{16}\,$g, which is a realistic amount of mass for a CME. The 3D simulation is compared with three 2.5D simulations having different values for $\rho_{\rm cme}$. The first axisymmetric simulation has the same initial parameters as the 3D simulation. Since in the axisymmetric case the initial plasma blob corresponds to a torus of higher density around the Sun, the total amount of mass in the torus is much higher than the total amount of mass in the 3D plasma blob, namely: $4\times10^{17}\,$g, which is higher than the observational values. The second 2.5D simulation has the same amount of mass in the torus as in the 3D plasma blob. Of course, it is not only the amount of mass in the blob that is important for the CME evolution, but also the initial momentum. In a third axisymmetric simulation, the parameters $d_{\rm cme}$ and $\rho_{\rm cme}$ are adapted in such a way that, if only a part of the torus would be considered, that has the same width as the 3D plasma bubble, then this part of the torus has the same initial volume and the same amount of initial momentum in the radial direction as the 3D plasma blob.

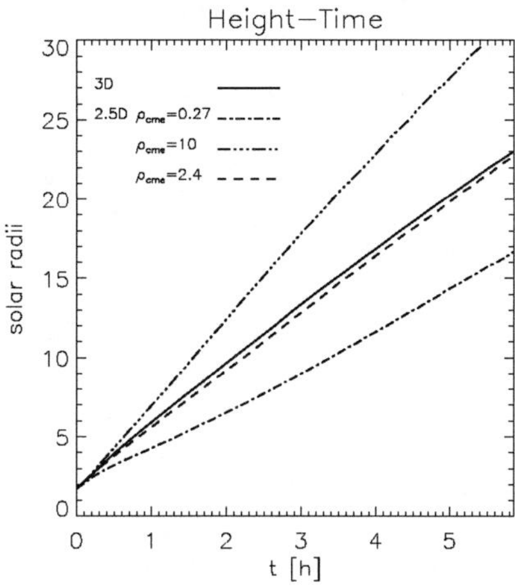

FIGURE 5. The position of the CME front along the equator versus time.

Figure 5 shows height-time plots of the CME front along the equatorial plane. From this plot it can be seen that the 2.5D simulation having the same amount of momentum compares best with the 3D result. It can be concluded that axisymmetric simulations can be used as a first approach to estimate the time of arrival of the shock and the density and velocity structure and that they resemble well the 3D result, provided that the appropriate initiation parameters are chosen. For more details about the simulation we refer to Jacobs et al. [13].

3D evolution of a flux rope

CMEs are often associated with erupting prominences and magnetic clouds. The solar magnetic field plays certainly a role in the CME triggering mechanism. White light images taken by coronagraphs often show the presence of a helical structure in the CME. In that sense, the previous simulations are clearly missing something essential, namely the presence of additional magnetic field in the plasma cloud. In this subsection the magnetised extension of the previous model is presented. This simulation is fully 3D and does not have a 2.5D counterpart.

Romashets and Vandas [14] derived analytical solutions of force free magnetic fields in toroidal geometry and applied these models to predict the profile of the magnetic field inside the a magnetic cloud. The magnetic field formulation of Romashets and Vandas [14] in term of Bessel functions is a modification of the force free solution of Miller and Turner [15] and reads:

$$\begin{aligned}
B_{r'} &= B_0 \frac{R_0 - 2r'\cos\theta'}{2\alpha R_0(R_0 + r'\cos\theta')} J_0(\alpha r')\sin\theta' \\
B_{\varphi'} &= B_0\left(1 - \frac{r'}{2R_0}\cos\theta'\right) J_0(\alpha r') \\
B_{\theta'} &= B_0 \frac{R_0 - 2r'\cos\theta'}{2\alpha R_0(R_0 + r'\cos\theta')} J_0(\alpha r')\cos\theta' - B_0\left(1 - \frac{r'}{2R_0}\cos\theta'\right) J_1(\alpha r')
\end{aligned}$$

where (r', φ', θ') are toroidally curved cylindrical coordinates, R_0 is the major radius of the torus, and B_0 is the strength of the toroidal component at $r' = 0$. The boundary of the torus is defined by $r' = r_0$, with r_0 the minor radius of the torus. The value of the constant α is such that αr_0 is the first root of the first order Bessel function J_0. This description of the magnetic field inside a torus does not show dependency to the toroidal direction, is divergence-free by definition, and is current free for large aspect ratio's ($R_0/r_0 \gg 1$). To this flux rope the stretching formula of Gibson and Low [16] is applied. The flux rope solution is subjected to the stretching transformation $r \rightarrow r - a$. For $a < 0$, this transformation stretches space inward. It effectively contracts a sphere of radius a, centred at the origin, to a point at the origin, and transforms the space outside this sphere symmetrically in all directions, with the contractive deformation diminishing at larger r. The deformed flux rope is then superposed on a background wind model. Typical for prominences is the strong shear they possess with respect to the underlying magnetic field and so the flux rope is rotated such that the toroidal field inside the flux rope makes an angle of 70° with the polarity inversion line of the global solar dipole field. The part of the flux rope in the northern hemisphere has a positive magnetic flux, like the background magnetic field, and the southern part has negative flux. The orientation of the magnetic field is taken such that the flux rope has a magnetised inverse topology. The flux rope is fully confined in the closed field line region of the overlying field and $B_0 = 1.44$ G. Inside the flux rope an additional density and velocity profile is posed. Both profiles show a dependency on the radius and on the toroidal direction inside the flux rope. The amount of mass added to the flux rope is 4×10^{15} g. Since it was in our aim to mimic a very fast CME event, the initial maximum velocity in the flux rope was set to 4000 km s^{-1}. The initial magnetic configuration is shown in Fig. 6.

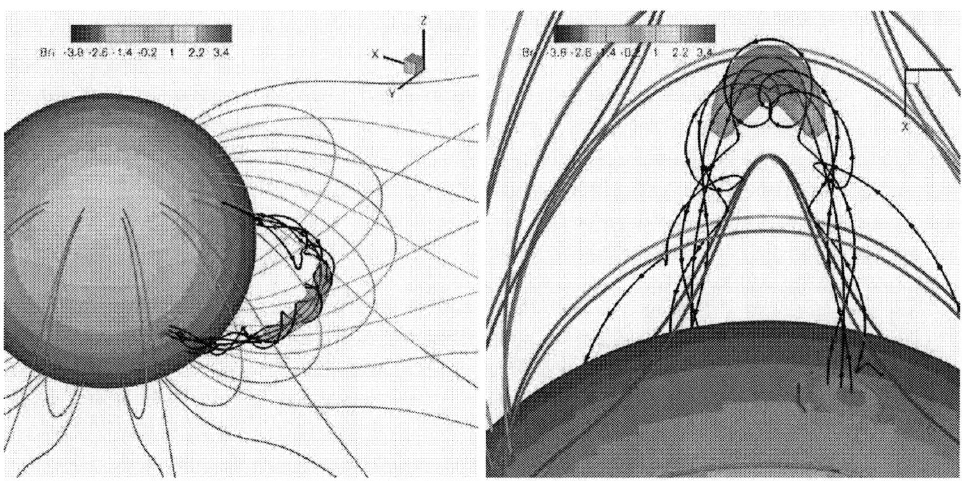

FIGURE 6. Side view (left) and zoom (right) of the initial magnetic field. Also shown is an iso-surface of the velocity ($v_r = 2000\,\mathrm{km\,s^{-1}}$). The solar surface is coloured with the radial magnetic field strength.

Reconstructed white light images are presented in Fig. 7. In the near future this simulation will be extended up to 1 AU and the results will be compared to flow variables observed in magnetic clouds by Earth orbiting satellites.

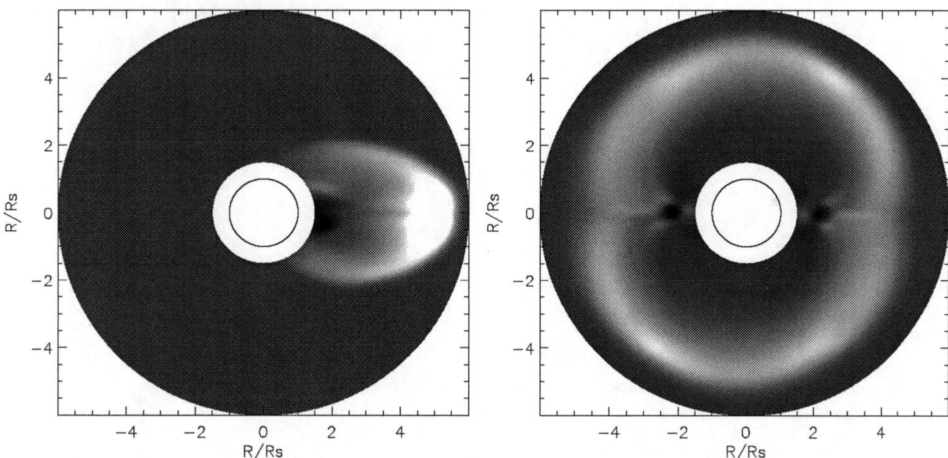

FIGURE 7. Reconstructed white light images showing the CME from different position angles. Left: Observer is in the equatorial plane and the CME is behind the limb; snapshot at 30min after onset. Right: frontal halo CME; snapshot at 1hrs 30min after onset.

CONCLUSION

We discussed the state-of-the-art CME initiation and IP evolution models at the CPA. First, we discussed the different solar wind models that are used as background for the CME simulations. Next, we briefly discussed CME initiation models based on magnetic foot point shearing and magnetic flux emergence, including a promising result on the break out model superposed on an MHD wind. In spite of extensive parameter studies, however, the shearing models are not able to produce the fast CMEs that create shock waves in the IP space and are, therefore, important for space weather. When taking into account the drag of the background wind, flux emergence seems to yield CME velocities up to 800 km/s. For the CME evolution studies we applied the much simpler, so-called 'density-driven', CME model. We then discussed recent fully 3D CME evolution simulations. The visualisation of these result in terms of white light images enables a direct comparison with LASCO observations. Such comparisons of theoretical results with observational data now become worth-while and are necessary for further improving the mathematical models of CMEs which are still too simplifying and not able to explain all aspects of initial and IP CME propagation.

ACKNOWLEDGMENTS

These results were obtained in the framework of the projects GOA/2004/01 (K.U.Leuven), G.0304.07 (FWO-Vlaanderen) and C 90196 and C 90203 (ESA Prodex 8). Financial support by the European Commission through the SOLAIRE Network (MTRN-CT-2006-035484) is gratefully acknowledged. The numerical results were obtained on the HPC cluster VIC of the K.U.Leuven.

REFERENCES

1. G. Tóth, *Astrophys. Lett. & Comm.* **34**, 245 (1996).
2. D. S. Balsara, and D. S. Spicer, *J. Comp. Phys.* **149**, 270 – 292 (1999).
3. C. P. T. Groth, D. L. De Zeeuw, T. I. Gombosi, and K. G. Powell, *J. Geophys. Res* **105**, 25053 – 25078 (2000).
4. W. Manchester, T. Gombosi, I. Roussev, D. De Zeeuw, I. Sokolov, K. Powell, G. Tóth, and M. Opher, *J. Geophys. Res* **109**, A01102 (2004).
5. S. A. Jacques, *ApJ* **226**, 632 – 649 (1978).
6. C. Jacobs, S. Poedts, B. van der Holst, and E. Chané, *A&A* **430**, 1099 – 1107 (2005).
7. E. Chané, C. Jacobs, B. van der Holst, S. Poedts, and D. Kimpe, *A&A* **432**, 331 – 339 (2005).
8. C. J. Schrijver, and M. L. DeRosa, *Sol. Phys.* **212**, 165–200 (2003).
9. B. van der Holst, and R. Keppens, *J. Comp. Phys.* (in press).
10. C. Jacobs, S. Poedts, and B. van der Holst, *A&A* **450**, 793 – 803 (2006).
11. G. Dubey, B. van der Holst, and S. Poedts, *A&A* **459**, 927–934 (2006).
12. P. MacNeice, S. K. Antiochos, A. Phillips, D. C. Spicer, C. R. DeVore, and K. Olson, *ApJ* **614**, 1028 – 1041 (2005).
13. C. Jacobs, B. van der Holst, and S. Poedts, *A&A* (2007), in press.
14. E. P. Romashets, and M. Vandas, "Interplanetary magnetic clouds of toroidal shapes," in *Proc. ISCS 2003 Symposium*, 2003, pp. 535 – 540.
15. G. Miller, and L. Turner, *Phys. Fluids* **24**, 363 (1981).
16. S. E. Gibson, and B. C. Low, *ApJ* **493**, 460 – 473 (1998).

Study of the interface between coronal holes and streamers

Giancarlo Noci*, Elena Gavryuseva†, Daniele Spadaro** and Roberto Susino‡

*Dipartimento di Astronomia e Scienza dello Spazio - Università di Firenze
†INAF - Osservatorio Astrofisico di Arcetri
**INAF - Osservatorio Astrofisico di Catania
‡Dipartimento di Fisica, Sezione Astrofisica - Università di Catania

Abstract. A technique to determine slow outflow speeds in the solar corona by means of the line ratio of the OVI resonance doublet was described in a previous paper [19]. The application of this technique to a streamer structure was also described in that paper. We extend here those results by analyzing two more streamers, both present at the west limb, one on May 4, 1996, the other on June 5, 1996. We find results which confirm the previous ones.

Keywords: EUV spectroscopy, solar corona, solar wind sources
PACS: 96.60.P-; 96.60.pf: 96.60.Vg; 96.50.Ci

INTRODUCTION

The SOHO observations (mainly from UVCS) have shown that the streamer plasma is different from the surrounding plasma, not only for its larger density and electron temperature, which was known, but also because it is characterized by temperature equilibrium, by a reduced abundance of oxygen (and of other ions) and by low or zero outflow speed [1], [2], [3], [4].

Furthermore, coronal holes and streamers are characterized by very different magnetic field topologies: open field lines in the coronal holes, closed ones in the streamers. This difference is probably the cause of all the other differences quoted above.

The study of the interface between these two magnetic structures is of great interest for coronal physics. It will probably lead to the understanding of a number of problems, as the origin of the slow solar wind and the mechanism of its acceleration. This is because the slow solar wind, which is known to be confined, in the interplanetary space, to a relatively narrow region around the current sheet, is widely believed to originate in magnetic tubes at the boundary between open field region and closed field region, which, at the cusp, may or may not go through the reconnection region.

To study the properties of the interface between closed (streamers) and open (coronal holes) magnetic regions, we believe particularly important to determine the outflow speed. It has been established in the seventies that the fast solar wind originates from coronal holes. In these structures small speeds (from less than 10 km/sec to a few times 10 km/sec) have been determined at the coronal base [5], [6], [7], much higher speeds ($>$ 130 km/sec) at 3 solar radii of heliocentric distance (at least for polar coronal holes) [1], [8], [9], [10], [11]. As for the streamers, the outflow speed (or limits to

CP934, *Flows, Boundaries, Interactions*
edited by C. Dumitrache, V. Mioc, and N. A. Popescu

it) has been determined by means of the intensity ratio of the OVI resonance doublet (UVCS/SOHO), or by the observation of moving features (LASCO/SOHO). The results obtained show the existence of (i) small velocities (consistent with zero) in the streamer core and legs [1], [12], [13], [14]; (ii) velocities raising to $\sim$ 100 km/sec at $r = 4 - 5R_{\odot}$ (where r is the heliocentric distance projected on the plane of the sky and $R_{\odot}$ the solar radius) [12], [9], while the same velocities at somewhat lower heliocentric distances were found in [15] and [14]; (iii) velocities reaching well beyond 100 km/sec at considerably larger heights [1], confirmed by the observation of blobs of plasma intermittently outflowing along the streamer axis above the cusp, at speeds rising to 300 - 400 km/sec at the heliocentric distance $r = 30R_{\odot}$ [16]; (iv) sharp gradient of speed at the streamer/coronal hole boundary [17], [18], [13].

A progress in this field would arise from the capability to determine speeds of the order of 20 - 30 km/sec. In the following we will show that it is possible to determine such small speeds in some streamers.

DIAGNOSTICS BASED ON THE OVI RESONANCE DOUBLET

We repeat here, for the sake of clarity, the arguments given in a previous work [19].

The total intensity of a resonance line in the solar corona is given by:

$$I = \int_{-\infty}^{\infty} [\int_{0}^{\infty} j'_{rad}(x,\nu)d\nu + j_{coll}(x)]dx, \tag{1}$$

where $j_{coll}(x)$ is the collisional emissivity, $\int_0^\infty j'_{rad}(x,\nu)d\nu$ the radiative one, and x is a coordinate along the line of sight. The radiative emissivity depends on the velocity, with respect to the source of radiation, of the volume element absorbing (and re-emitting) the radiation, owing to the Doppler effect. If the radiation emitted by the source consists of an emission line, the relative velocity causes its center to move away from the center of the absorbing profile, in the frame of the absorbing ion. The consequence is a reduced absorption and therefore a reduced re-emission (Doppler dimming). The emission could go to zero for sufficiently high speed. If the spectrum of the source of exciting radiation contains a second emission line on the short wavelength side of the line considered, its shift can cause the coronal absorption profile to overlap the second line (Doppler pumping). [3, and references therein.]

In the case of the OVI ions, the ground configuration is $1s^2 2s$, which gives rise to the resonance transitions $1s^2 2s(^2S_{1/2}) - 1s^2 2p(^2P_{1/2})$ (λ1037.6) and $1s^2 2s(^2S_{1/2}) - 1s^2 2p(^2P_{3/2})$ (λ1032). These two lines, in the solar corona, can be excited by collisions with the free electrons or by photon absorption.

Accordingly, if we apply eq.(1) to the 1037.6 Å line of OVI, we can write, in short,

$$I = C + R + P, \tag{2}$$

where C is the collisional term, R is the part of the radiative term arising from excitation due to the transition region radiation in the 1037.6 line and P the part due to the radiation from neighboring lines, such as the chromospheric lines 1037.0 and 1036.3 of CII (pumping term). As for the 1032 line of this ion, its collisional term is two times that of the 1037 line, while the radiative term is four times as large [20]. The pumping term is zero for this line.

Hence the line ratio is given by:

$$p = \frac{I_{1037}}{I_{1032}} = \frac{1+R/C}{2+4R/C} + \frac{P/C}{2+4R/C}, \tag{3}$$

which shows that the intensity ratio can be any positive number lager than 0.25. However, if P=0 the intensity ratio is between 0.25 and 0.5. This equation shows clearly that a single determination of p in the interval 0.25 - 0.5 does not give, by itself, any information on the speed but only on the ratio R/C. The speed could be obtained only if the electron density and the intensity of the exciting radiation were known. Information on the speed (lower limit) is obtained if the intensity ratio is larger than 0.5, because this means $P > 0$, which occurs if $v \gtrsim 100$ km/sec in the coronal regions, along the line of sight, which contribute most to the intensity [21].

However, it is possible to obtain some important information on the outflow speed if one considers an entire region. Let us consider the behaviour of the intensity ratio as a function of the height. In a static atmosphere, C, which depends on the square of the density, decreases much more rapidly than R, which depends on the density times the solid angle subtended by the solar disk. Hence the ratio R/C increases and p decreases toward the value 0.25. However, if significant outflow is present at some height, R decreases rapidly owing to Doppler dimming, and may cause the intensity ratio to increase back toward the value 0.5. If the outflow speed increases with height the behaviour of the intensity ratio is as shown schematically in fig. 1.

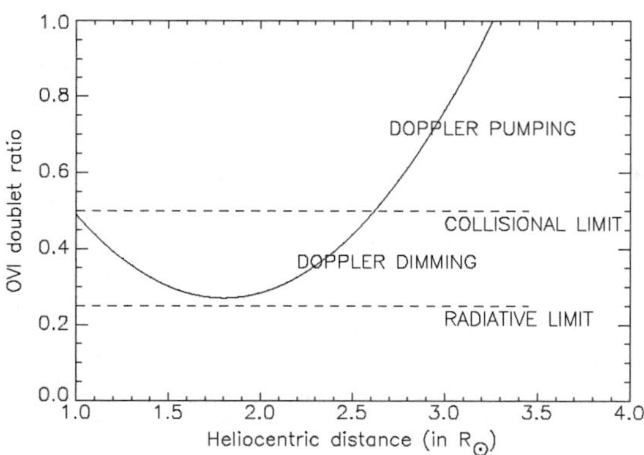

FIGURE 1. *Schematic behavior of the line ratio of the OVI resonance doublet in the solar corona.*

Once it has become significant, Doppler dimming increases rapidly as the velocity increases. Therefore the minimum of the curve in fig. 1 corresponds to the beginning of a significant Doppler dimming, hence, roughly, to $v = v_m = 20$ km/sec (i.e. half the e-1 half-width of the velocity distribution function of the OVI ions at $T = 1.5 \times 10^6$K). The identification of this point is, indeed, the identification of the point where the outflow speed reaches a value close to 20 km/sec.

SLOW OUTFLOW SPEED AROUND STREAMERS

Since the standard UVCS observations are for projected heliocentric distances larger than $r = 1.5R_\odot$, they do not cover, as far as we are aware, the $v = v_m$ curve in coronal holes, where it lies at lower heliocentric distances.

Therefore we have turned our attention to the streamer regions. It appears that there are differences between streamers in the phase of activity minimum and streamers of the active sun [22], [23],[24],[25],[26], differences wich may affect also the solar wind sources [27],[28]. At the origin of these observed differences there is, most probably, the difference in the underlying photospheric magnetic field, having a much more complex topology in the activity phase. Hence, the easier situation to analyze is probably the one involving streamers at activity minimum. Owing to this, we have studied, for the moment, 'quiescent' minimum phase streamers. Among these, the streamers characterized by a dark core in their OVI images, which have been observed by the UVCS/SOHO instrument at the last activity minimum [1],[29].

We have analyzed, in a previous study [19], a streamer structure present at the West solar limb on 25 - 28 June, 1996. We present, here, the results of a similar analysis on two more streamers observed by UVCS: one on June 6, 1996 (synoptic observations), one on May 4, 1996 (special observations). Both at the west limb, with the slit perpendicular to the equator; the slit projected heliocentric distances (in solar radii) were 1.5 1.7 1.9 2.2 2.6 3.0 (June 6), and 1.4 1.45 1.5 1.6 1.7 1.85 2.0 2.25 2.5 3.0 3.5 (May 4).

For these streamers we have calculated, for each bin in the spatial direction (21 arcsec for June 5, 28 arcsec for May 4), the positions of the minima of the $p(r)$ curve (eq.(3)), and then averaged the sequence of minimum positions over six (June 5) or five (May 4) spatial bins by a running mean. Since, when the lowest value of the intensity ratio, for a given position, is at the heliocentric distance upper limit this value is taken as the minimum, the portion of the running mean curve which is interested by these values is marked differently from the rest (figs. 2 and 3). Such portions, clearly, represent lower limits for the heliocentric distance of the minima of the intensity ratio.

The statistical analysis of the central part of the curves $p(r)$, i. e. the one around the streamer axis, has not been completed yet; therefore, we discuss, here, the lateral sides of these curves, which represent, according to the previous section, the line where the speed of the outflowing plasma becomes significant ($\simeq$ 20 km/sec).

Similarly to the analysis reported in [19], we have studied the statistical significance of these curves by calculating the functions $p(r)$ with intensity data averaged over six (June 5) or five (May 4) spatial bins, for four interval in the spatial directions chosen as in [19]: strips 1 and 5 are on the lateral sides of the streamer and strips 2 and 4 internal to them but still away from the streamer axis (they overlay the intensity maxima of the 1032 line at $r = 3R_\odot$).

The resulting intensity ratios are given in fig. 4, as a function of the heliocentric distance, the error bars representing the statistical errors (standard deviations of the mean value). This figure shows that, both for May 4 and June 5, no significant minimum is present for strips 2 and 4 up to the highest value of heliocentric distance observed. Hence this statistical analysis is consistent with the absence of Doppler dimming below that limit in correspondence of the two lateral brightness maxima of OVI emission. On the contrary, a minimum of the intensity ratio below or around $r \sim 2R_\odot$ is clearly present

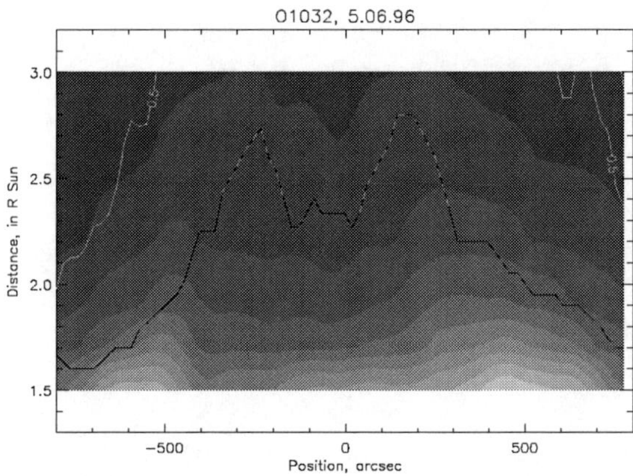

FIGURE 2. *Image of the west limb streamer on June 5, 1996 in the OVI λ 1032 radiation. The image results from an interpolation between strips of the corona corresponding to the UVCS entrance slit. The x axis is perpendicular to the equatorial plane on the plane of the sky, 0 representing the equator; the y axis is the heliocentric distance on the equatorial plane. Contour plots: isobrightness curves; black and dashed curve: positions of the minimum of the OVI intensity ratio (running mean curve): dashed portions: lower limits to the heliocentric distance of the $v \simeq 20$ km/sec curve; black portions: heliocentric distance of the $v \simeq 20$ km/sec curve (see text); white curve: positions where the OVI intensity ratio has the value 0.5 ($v \simeq 100$ km/sec).*

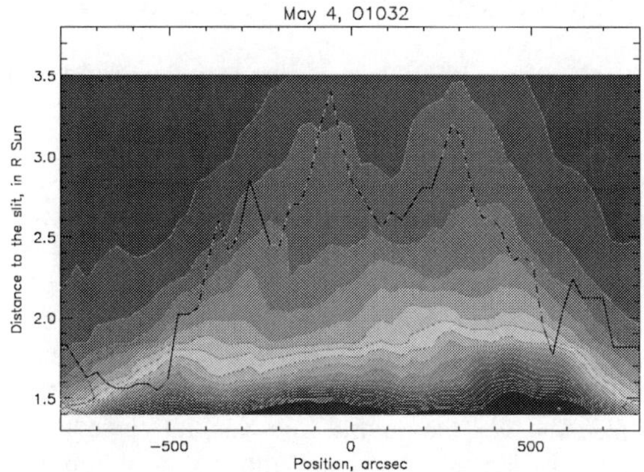

FIGURE 3. *Same as fig. 2 for May 4, 1996.*

in strips 1 and 5, confirming that a significant outflow speed is present at that heliocentric distance at the lateral hedges of the streamers.

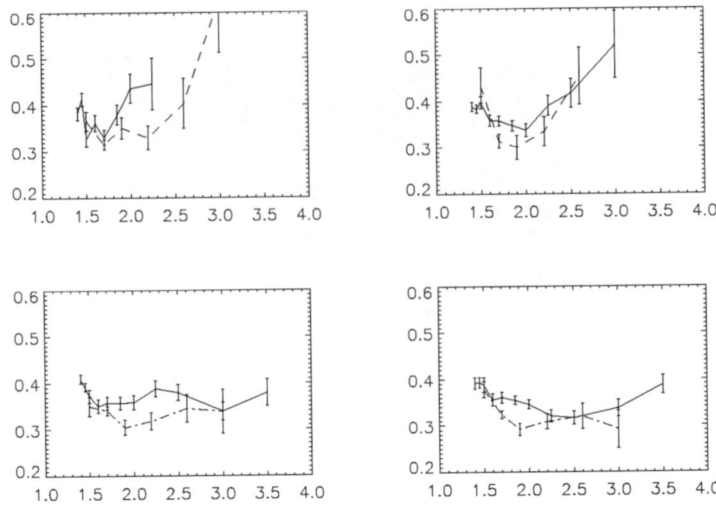

FIGURE 4. *a) OVI intensity ratios for May 4 (solid line) and June 5, 1996 (dash or dash-dotted) at the west limb in strips 1 and 5 (upper panels); 2 and 4 (lower panels).*

PHOTOSPHERIC MAGNETIC FIELD

There is no doubt that the magnetic field structure is at the origin of the characteristics of the streamers. Hence, for the streamers studied here, we examine the photospheric field, which is given in figs. 5 and 6. These figures show, for each day given on the abscissa, the photospheric magnetic field at central meridian passage (CMP). Hence for May 4, the magnetic field presumably present at the west limb is that corresponding to April 27-28 in fig. 5, and, for June 5, that corresponding to May 29-30 in fig. 6.

In the period April - May 1996 the photospheric field did not have a simple bipolar structure. An active region was present at low latitudes, with CMP during 11 - 15 May. In the previous CMP of this longitude the active region had not developed yet, only a hint of it being present. No other active region was present, as one would expect in a period of activity minimum. However, the neutral line was not close to the equator, running at around -40 -50 degrees of latitude, but many intrusions of southern polarity existed at lower (in absolute value) latitudes and also across and above the equator. We did not make calculations of the coronal field in the temporal intervals of our analysis, however it seems to us that the photospheric field was compatible with a quadrupolar structure of the coronal field. We are not able to establish, at present, if this structure is a necessary condition for the presence of a dark core in the OVI streamer images, as observed in the streamers studied.

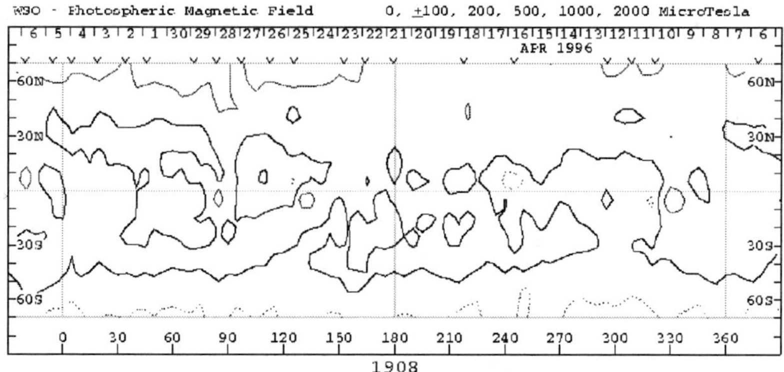

FIGURE 5. *Photospheric magnetic field on April-May 1996 (From Wilcox Solar Observatory). The black line is the neutral line. The steps of higher magnetic field values (northern or southern polarity) are indicated by gray or dotted contours.*

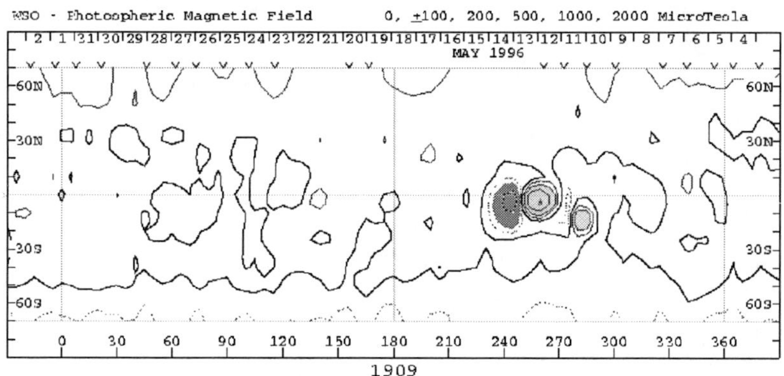

FIGURE 6. *Photospheric magnetic field on May-June 1996 (From Wilcox Solar Observatory). For explanations see caption of fig. 5.*

CONCLUSIONS

We have extended the analysis of [19] to two more streamers, which present, as those analyzed previously, a dark core in the OVI images, in contrast with those in Ly-alpha, a phenomenon which has been interpreted as due to oxygen abundance depletion [2], [3]. Having not yet completed the analysis of the central parts of the streamers studied here, we remain with this interpretation.

These new data confirm the possibility to identify the streamer 'lateral border' as the one where the outflow speed of the coronal plasma reaches a value of ~ 20 km/sec. This border is very close to where the drop in intensity, from the streamer to the adjoining coronal hole, takes place.

Aknowledgments Thi work was supported in part by Agenzia Spaziale Italiana, through contract I/035/05/0 to Istituto Nazionale di Astrofisica.

REFERENCES

1. J. L. Kohl, G. Noci, E. Antonucci et al., *Solar Phys.* **175**, 613-644 (1997).
2. G. Noci, J. L. Kohl, E. Antonucci et al., in *The Corona and Solar Wind near Minimum Activity (SOHO-5)*, edited by O. Kjeldseth-Moe, ESA SP-404, pp. 75-84 (1997).
3. J. C. Raymond, J. L. Kohl, G. Noci et al. *Solar Phys.* **175**, 645-665 (1997).
4. J. L. Kohl, G. Noci, S. R. Cranmer and J. C. Raymond, *A&A Rev.* **13**, 31-157 (2006).
5. D. M. Hassler, I. E. Dammasch, P. Lemaire et al., *Science* **283**, 810-813(1999).
6. L. Teriaca, G. Poletto, M. Romoli, D. A. Biesecker, *ApJ* **588**, 566-577 (2003).
7. S. Patsourakos and J.-C. Vial, *A&A* **359**, L1-L4 (2000).
8. S. R. Cranmer, J. L. Kohl, G. Noci et al., *ApJ* **511**, 481-501 (1999).
9. G. Poletto, S. T. Suess, D. A. Biesecker et al., *JGR* **107**, SSH 9-1 (2002).
10. L. Zangrilli, G. Poletto, P. Nicolosi et al., *ApJ* **574**, 477-494 (2002).
11. E. Antonucci, M. A. Dodero, S. Giordano et al., *A&A* **416**, 749-758 (2004).
12. L. Strachan, R. Suleiman, A. V. Panasyuk et al., *ApJ* **571**, 1008-1014 (2002).
13. D. Spadaro, R. Ventura, G. Cimino, M. Romoli, *A&A* **429**, 353-360 (2005).
14. E. Antonucci, L. Abbo, M. A. Dodero, *A&A* **435**, 699-711 (2005).
15. L. Abbo and E. Antonucci, in *The Streamer Boundary and the Source of the Slow Wind (SOHO-11)*, ESA SP-508, PP. 477- (2002).
16. N. R. Sheeley, Jr., Y.-M. Wang, S. H. Hawley et al., *ApJ* **484**, 472-478 (1997).
17. E. Antonucci, S. Giordano, C. Benna et al., in *The Corona and Solar Wind near Minimum Activity (SOHO-5)*, edited by O. Kjeldseth-Moe, ESA SP-404, pp. 175-182 (1997).
18. S. R. Habbal, R. Woo, S. Fineschi et al., *ApJ* **489**, L103-L106 (1997).
19. G. Noci and E. Gavryuseva, *ApJ* **658**, L63-L66 (2007).
20. J. L. Kohl and G. L. Withbroe, *ApJ* **256**, 263-270 (1982).
21. G. Noci, J. L. Kohl and G. L. Withbroe, *ApJ* **315**, 706-715 (1987).
22. R. A. Frazin, A. Ciaravella, E. Dennis et al., *Space Sci Rev.* **87**, 189-192 (1999).
23. C. R. Foley, S. Patsourakos, J. L. Culhane, D. MacKay, *A&A* **381**, 1049-1058 (2002).
24. M. Uzzo, Y.-K. Ko, J. C. Raymond et al. *ApJ* **585**, 1062-1072 (2003).
25. L. Strachan, M. Baham, M. P. Miralles, A. V. Panasyuk, in *UVCS/SOHO Measurements of Heathing in Coronal Streamers (SOHO-15*, edited by R. W. Walsh, J. Ireland, D. Danesy, B. Fleck, ESA SP-575, pp. 148-153 (2004).
26. R. Ventura, D. Spadaro, G. Cimino, M. Romoli, *A&A* **430**, 701-712 (2005).
27. P. Hick, B. V. Jackson, S. Rappoport et al. *GRL* **22**, 643-646 (1995).
28. P. C. Liewer, M. Neugebauer and T. Zurbuchen, *Solar Phys.* **223**, 209-229 (2004).
29. G. Noci, J. L. Kohl, E. Antonucci et al., *Adv. Space Res.* **20**, 2219-2230 (1997).

Latitudinal and Radial Variation of Solar Corona Rotation at Solar Minimum

S. Giordano*, S. Mancuso* and M. Romoli†

*INAF - Astronomical Observatory of Torino, Pino Torinese, Italy
†Dip. Astronomia e Scienza dello Spazio, Universitá di Firenze, Firenze, Italy

Abstract. The rotation of the solar corona at different heliolatitudes from 1.5 to 3.0 $R_\odot$ from Sun center has been studied at solar minimum from the reconstructed intensity time series of the O VI 1032 Å and H I Lyα 1216 Å spectral lines and visible light polarized brightness obtained by the observations of UVCS/SOHO instrument. The time period analyzed range from mid May 1996 to mid May 1997, when, at solar minimum, some features persist for several rotations, thus allowing to analyze the UV and visible emission as time series modulated at the period of the solar rotation. The coronal differential rotation rate significantly differs from that of the photospheric plasma. The estimated equatorial synodic rotation period of the corona at 1.5 $R_\odot$ is 27.48 $\pm$ 0.15 days. The study of the latitudinal variation shows that the UV corona decelerates towards the photospheric rates from the equator up to the poleward boundary of the mid–latitude streamers, reaching a peak of 28.16 $\pm$ 0.20 days around $+30°$ from the equator at 1.5 $R_\odot$, while a less evident peak is observed in the northern hemisphere, suggesting a real north–south rotational asymmetry, the northern hemisphere the rotation looks more solid-body-like and slower than in the southern hemisphere. The mid-latitude results are also confirmed by the visible light data available at 1.75 and 2.0 $R_\odot$. The study of the radial rotation profiles shows that the corona is rotating almost rigidly with height, but we find an abrupt increase by about half a days between 2.3 and 2.5 $R_\odot$. The larger radial and latitudinal gradients of the rotation rates are localized at the boundary between the open and closed field lines, suggesting that in these regions the differential rotation might be a source of magnetic stress and, consequently, of energy release.

Keywords: Sun, corona, rotation, spectral lines
PACS: 96.60.P, 96.60.Bn

INTRODUCTION

The studies of solar corona rotation are of particular importance in that there are still several open questions on how the photospheric differential rotation is transferred into the corona and on the rotation rate of different small–scale coronal features. The differential rotation of the corona is observed to be much less pronounced than in the photosphere, in particular, away from the equator, the corona tends to rotate faster and more rigidly than photospheric plasma. The investigation of the rotation rate of the solar corona is especially interesting because the interaction between the differentially-rotating photosphere and the near-rigid rotating corona could be a possible energy source for both coronal heating and solar wind acceleration and, consequently, may provide a powerful constraint to solar wind models. However, at present, there have been no rotation studies that make extensive use of ultraviolet (UV) emission lines.

Previous investigations have mainly focused their analyses on observations of the coronal rotation in white–light (e.g. [1]), Fe XIV *green line* ([2]), microwave ([3]), X-

CP934, *Flows, Boundaries, Interactions*
edited by C. Dumitrache, V. Mioc, and N. A. Popescu

ray ([4]), and through extrapolation of photospheric data (e.g. [5]). More recently, using data from the Large Angle Spectrometric Coronagraph (LASCO) aboard the Solar and Heliospheric Observatory (SOHO) spacecraft, the rotation of the inner and outer corona at solar minimum has been investigated by [6] and [7].

In the UV, coronal rotation rates can be inferred thanks to the regular occurrence of localized coronal features that, being much brighter than the background corona, act as *tracers* whenever they have sufficient stability to reappear at the same limb for several rotations. The measurement of the rotation of the extended corona can thus be achieved by using regular synoptic observations obtained with the Ultraviolet Coronagraph Spectrometer (UVCS) instrument on SOHO, which is able to provide long and uninterrupted time series of data ranging from about 1.5 to 3.0 $R_\odot$ from Sun center.

The latitudinal and radial dependence of the coronal rotation rate during solar minimum conditions are presented in this paper, the results are based on the analysis of the available daily ultraviolet and visible observations made with the UVCS instrument from mid May 1996 to mid May 1997. The latitudinal range of our data extends from the solar equator to about 30° north and south respect to the poles, with enough spatial resolution, in the UV, to search for the latitudinal dependence of the rotation rates. Moreover the data from the UVCS visible light channel are also used as an independent determination of the rotation periods at the mid–latitudes.

OBSERVATIONS AND DATA REDUCTION

The UVCS instrument aboard SOHO is an internally and externally occulted coronagraph, consisting of two spectrometric channels for the observation of spectral lines in the UV range and a White Light Channel (WLC) for polarimetric measurements of the extended solar corona. The UVCS slit can be moved along the radial direction to observe the solar corona between 1.4 and 10 $R_\odot$ with a field of view of 40'. In order to cover all the possible position angles, the slit can be rotated by 360° about an axis pointing to the center of the Sun. The WLC of UVCS is a coronagraph polarimeter operating in the wavelength range from 4500 to 6000 Å, it measures the polarized brightness (pB) of the corona from 1.5 to 5 $R_\odot$, with a spatial resolution of $14'' \times 14''$, in one single point at each time, which is coregistered with the field of view of the ultraviolet spectrometers of UVCS ([8]).

In this work, we present a coronal rotation study that makes use of observations of the O VI and H I Lyα spectral lines at 1032 Å and 1216 Å, respectively, and of the visible light data. We focus our investigation on the period of minimum of solar activity, when the solar corona is relatively stable and coronal structures are sufficiently long lived. In particular, we selected a one year observation period, from mid May 1996 to mid May 1997. In the time interval of interest for this work, UVCS run an anti–clockwise daily synoptic observation program which covered the full corona from 1.5 to 3.0 $R_\odot$ at eight different roll angles separated by an angular step of 45°. On average, the cadence of the data was about one per day, unevenly spaced in time. The instrument rolls analyzed in this work were those positioned at the mid–latitude regions (45°, 135°, 225°, and 315°), and at the equatorial regions (90° and 270°). The slit distances from the Sun center were 1.5, 1.75, and 2.0 $R_\odot$ for the mid–latitude regions, and 1.5, 1.7, 1.9, 2.2, 2.6, and

3.0 $R_\odot$ for the equatorial regions. Equatorial distances higher than 2.0 $R_\odot$ have only been analyzed to study the radial variation of the rotation rate, since the low count rates at these heliocentric distances did not allow a latitudinal variation study.

All the subsequent UV exposures with the same mirror pointing were summed up together, so that the average exposure time of the time series elements is 670 seconds. For each selected roll angle, we used the data acquired in a $\pm 850''$ range around the center of the slit and, in order to increase the statistics, we further rebinned the data to 21 spatial elements along the slit for each observation, thus reducing the spatial resolution along the slit to $\simeq 81''$, which corresponds to a latitude step of about 3.2° at 1.5 $R_\odot$, and 2.4° at 2.0 $R_\odot$. We restricted our analysis to a latitudinal range of about ±60° around the equator, where the signal to noise ratio was found to be high enough for a reliable period determination. We should emphasize that the observed distances (projected into the plane of the sky) are not constant along the slit. For example, for a nominal mirror pointing at 1.50 $R_\odot$, in the selected $\pm 850''$ range around the center of the slit, the actual observed distance ranges from 1.48 ± 0.02 $R_\odot$ at the slit center to as much as 1.70 ± 0.02 $R_\odot$ at the edges of the slit. All the UV data in the period under study were collected and used to build the synoptic maps at a fixed mirror pointing by fitting the selected spectral lines to obtain the total intensity from each coronal region element.

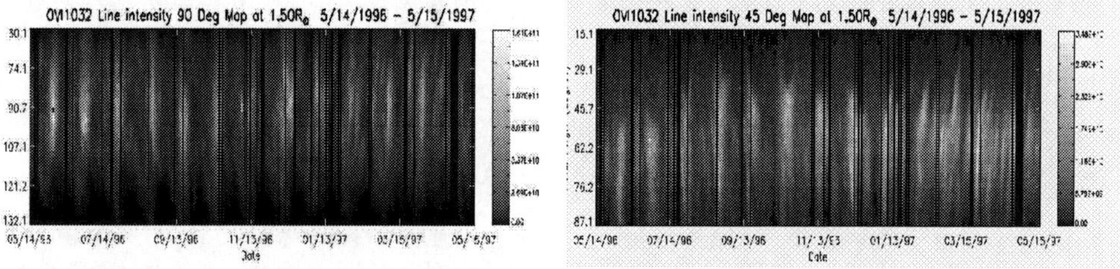

FIGURE 1. O VI 1032 Å intensity synoptic maps in the time interval from mid May 1996 to mid May 1997 at with instrumental slit at 90° (left) and 45° (right) anti–clockwise from the north pole at 1.50 $R_\odot$ heliocentric distance. The intensities are measured in *photons* cm^{-2} s^{-1} sr^{-1}.

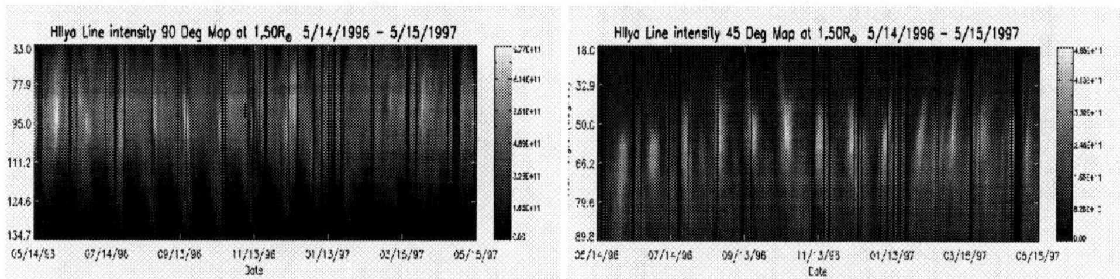

FIGURE 2. H I Lyα 1216 Å intensity synoptic maps as previous figure

In Figure 1, we show the O VI 1032 Å intensity synoptic maps obtained with the UVCS mirror pointed at 1.5 $R_\odot$in the equatorial region (right panel) and in the north–east mid–latitude region (left panel). Similar maps have been built at all the analyzed roll and heights. Indeed, the intensity map of Figure 1 shows a clear modulation which can be readily attributed to the rotation of persistent features through several consecutive rotations. Previous studies (e.g., [1]) showed that large–scale features are actually seen to persist for several synodic rotations, a trend that is also confirmed by our correlation study of the coronal structure lifetimes. Consequently, the intensity at a fixed location in the synoptic O VI 1032 Å image gives a suitable time series for analysis period. From the H I Lyα data we reconstruct the synoptic maps presented in Figure 2. The equatorial image (left panel) shows a diffuse and constant signal from the streamer belt, since these data were found less reliable for a thorough coronal rotation period study in the equatorial region, they will only be used for validating the rotation period results at mid–latitudes, where the signal is more periodic (see right panel of Figure 2).

The polarimeter consists of a rotating half wave retarder plate (HWRP) and a fixed linear polarizer. From the photon count rates at the three HWRP rotated positions, which are separated by 30° one from each other we derive the polarization brightness (pB) in a single point relative to the roll angle and to the heliocentric height of the instrument pointing. Finally, pB is normalized to the Sun center brightness. Due to an offset in the pointing of the WLC ($\simeq 231''$), the heliocentric heights of visible observation are slightly higher than nominal (from 0.01 to 0.02 $R_\odot$), and the actual position angle is increased in average by 7° with respect to the nominal roll. The WLC rolls presented in this work were those positioned at the mid–latitude regions and the slit distances from the Sun center were 1.75, and 2.0 $R_\odot$. The equatorial regions have been also analyzed but the pB time series shows a behaviour similar to the H I Lyα, therefore we use only the mid–latitudes data.

PERIOD DETERMINATION METHODOLOGY

Different time series analysis methods can be applied to determine the data periodicity. We used the Lomb–Scargle periodogram technique ([9]; [10]) as our main investigation tool for the analysis of the power spectrum of the O VI 1032 and H I Lyα intensity and pB time series for the selected time window, since it is able to deal with unevenly spaced data. In order to validate our results and obtain an independent period determination, we also used the autocorrelation analysis technique, applied over data interpolated onto an equally spaced grid.

For a given position along the slit, s, and observation time, t_i, which corresponds to the middle of the integration time interval, if we denote the intensity time series by $y_s(t_i)$ for $i = 1,...,N$, where N is the number of observations, the mean and the variance of the data are given, respectively, by

$$\bar{y}_s = \frac{1}{N}\sum_{i=1}^{N} y_s(t_i), \tag{1}$$

and

$$\sigma_s^2 = \frac{1}{N-1}\sum_{i=1}^{N} (y_s(t_i) - \bar{y}_s)^2 . \tag{2}$$

Unlike Fourier analysis, in which the Fourier frequencies are used, we assume that there are M test frequencies f_1, f_2, ..., f_M, with corresponding angular frequencies

$\omega_j = 2\pi f_j$, and periods $T_j = 1/f_j$, for $j = 1,...,M$. The Lomb–Scargle periodogram, normalized by the total variance of the data σ_s^2, is defined in [11] as

$$P_s(\omega_j) = \frac{1}{2\sigma_s^2}\left\{\frac{\left(\sum_{i=1}^N [y_s(t_i)-\bar{y}_s]\, cos\left[\omega_j(t_i-\tau_j)\right]\right)^2}{\sum_{i=1}^N cos^2\left[\omega_j(t_i-\tau_j)\right]} + \frac{\left(\sum_{i=1}^N [y_s(t_i)-\bar{y}_s]\, sin\left[\omega_j(t_i-\tau_j)\right]\right)^2}{\sum_{i=1}^N sin^2\left[\omega_j(t_i-\tau_j)\right]}\right\} \quad (3)$$

where τ_j is defined by

$$\tan(2\omega_j\tau_j) = \frac{\sum_{i=1}^N \sin(2\omega_j t_i)}{\sum_{i=1}^N \cos(2\omega_j t_i)} \quad (4)$$

The choice of M depends on the number of independent frequencies N_0. [12] investigate the relationship between M and N_0, giving a simple least squares formula to estimate N_0 from the number of observations N in a time series:

$$N_0 \approx -6.362 + 1.193N + 0.00098N^2. \quad (5)$$

This empirical formula is adequate for most purposes by actually taking M as N_0 ([13]). In our case, this choice is supposed to be valid also because there is no important data clumping. In fact we have approximately one measurement per day and rarely more than one measurement in a single day. This formulation is exactly equivalent to a linear least–squares fit of a sinusoid of given frequency ω to the time series data. For a detailed description of the Lomb–Scargle normalized periodogram properties see [10].

The frequency analysis was performed on the natural logarithm of the UV intensity data in order to compress their dynamic range and reduce the impulsive effects on the time series due to the recurrence of active region streamers. In order to remove possible long–term variations or trends due to the changing Sun–SOHO distance or uncorrected instrumental effects, the time series analyzed were detrended by subtracting a linear fit to the intensity data. In Figure 3 (left panel), we report an example of the O VI 1032 intensity time series observed in the east equatorial region (90° anti–clockwise from the north pole) at 1.5 $R_\odot$, and the relative power spectrum from the Lomb–Scargle periodogram analysis (lower left panel of Figure 3). The right panel of Figure 3 shows a pB time series obtained from the WLC data at 1.75 $R_\odot$ and 232°. The uncertainty on the frequency determination was evaluated following the [12] formulation:

$$\sigma_f = \frac{3\sigma_n}{4\sqrt{N}TA} \ [\mathrm{s}^{-1}], \quad (6)$$

where σ_n^2 is the variance of the noise after subtracting the least–squares fitted sine curve, T is the total length of the observing interval, and A is the amplitude of the signal obtained from a least–squares fit to the data of a sinusoidal curve of the peak frequency. We added a systematic error to the above uncertainty due to the frequency sampling step. The systematic errors on the period determination are $\sigma_{T,sys} \simeq 0.15$ days. Although we treated the east and west limbs independently, we finally performed a weighted average of the rotation periods from both limbs, with errors estimated by means of weighted standard deviations.

As extensively applied in previous studies of coronal rotation (e.g., [14]; [15]; [6]), we also derived an estimate of the mean synodic rotation period of the UV corona by using the autocorrelation technique over the same data set.

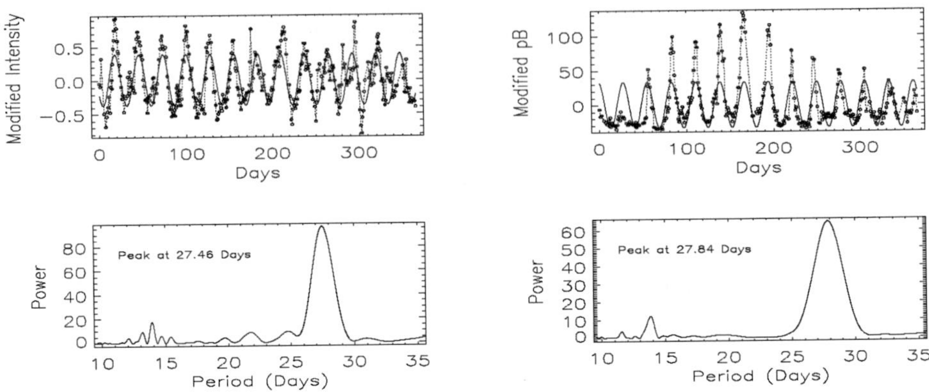

FIGURE 3. The upper-left plot shows the O VI 1032 modified intensity time series at 90° from the north pole at the heliocentric distance of 1.5 $R_\odot$, and, superimposed, the computed sinusoidal signal with a period of 27.46 days determined from the main peak of the power spectrum showed in the lower-left plot. The upper-right plot shows the visible light pB modified time series at 232° from the north pole (which corresponds to 128° in the east quadrant) at the heliocentric distance of 1.75 $R_\odot$, and, superimposed, the computed sinusoidal signal with a period of 27.84 days determined from the main peak of the power spectrum showed in the lower-right plot.

RESULTS

Latitudinal variation

By analyzing the O VI 1032 spectral line time series with the instrumental slit height ranging from 1.5 through 2.0 $R_\odot$, covering a latitudinal region of about ± 60° from the equator, we determine the latitudinal variations of the rotation periods estimated both from the Lomb–Scargle periodogram and the autocorrelation techniques. Up to the uncertainties of both the autocorrelation and the Lomb–Scargle analysis (about 0.15-0.30 days per rotation period near the equator), the results from the two techniques applied to the same data set agree remarkably, both qualitatively and quantitatively ([16]). By averaging over a spatial region of ± 7° around the equator, our best estimate for the equatorial rotation period of the corona at 1.50 $R_\odot$ is 27.48 ± 0.15 days with Lomb–Scargle periodogram technique and 27.44 ± 0.13 days with the autocorrelation technique, in agreement with the ± 45° latitude averaged coronal rotation determined by [6]. The variation of the coronal rotation with latitude at 1.5 $R_\odot$ (upper-left panel of Figure 4) shows that the UV extended corona increases its rotation period from the equator up to the poleward boundary of the mid–latitude streamers. The rotation period reaches a peak of 28.16 ± 0.20 days around 120° in the southern quadrant while, around 60° in the northern quadrant, at the symmetrical latitude with respect to the equator, there is a less evident peak at 27.57 ± 0.22 days. Over these peaks, towards the poles, the rotation period is seen to decrease again. At these higher latitudes, however, the period estimate is found to be difficult due to both low signal to noise ratios and projection effects of lower latitude features out of the plane of the sky. This departure from rigid rotation is in agreement with the results of [7] obtained using LASCO C1 data in the lower corona.

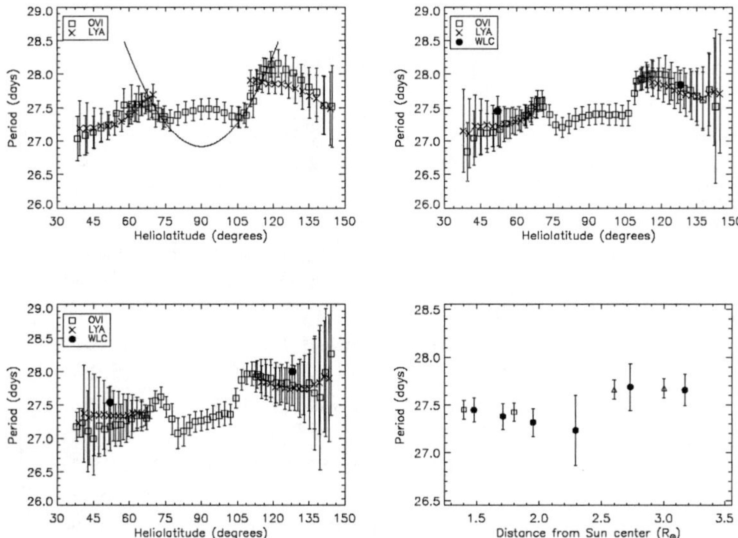

FIGURE 4. Latitude dependence of the coronal rotation rate between 1.5 and 2.0 $R_\odot$ obtained with the Lomb–Scargle periodogram technique from the analysis of O VI 1032 Å (open squares), H I Lyα 1216 Å (crosses) and pB (full dots) time series. The slit heliocentric distances are 1.50 $R_\odot$ (upper-left), 1.70 and 1.75 $R_\odot$ (upper-right) and 1.90 and 2.00 $R_\odot$ (lower-left). The solid line on the upper-left plot shows the differential rotation period for recurrent sunspot groups near the solar minimum obtained by [17]. The lower right plot shows the radial dependence of the rotation period around the solar equator. The full dots indicate the results from UVCS data while the open squares and triangles denote the period estimates from LASCO C1 and C2 observations, respectively ([6]).

The observed latitudinal variation of the rotation period is also confirmed by results obtained from the analysis of the intensity time series of the H I Lyα 1216 Å spectral line. As shown in Figure 4, the period estimates from these two independent time series show a good agreement in the mid–latitudes where the rotational modulation is expected to be stronger than in the bright and almost time–independent equatorial streamer belt at solar minimum. In fact, in the equatorial region, the hydrogen scattering across the magnetic lines is large (see Figure 2), and the period determination is affected by large errors and low periodicity significance. For these reasons, only the mid–latitude periods from H I Lyα data are presented. Although the overall coronal rotation is definitely more rigid with respect to the photospheric, it appears to match quite well the sunspot rotation in both hemispheres in a region around $\pm$ 15° $\div$ 25° from the equator while, closer to the equator, it appears to rotate slower than the sunspots (see upper-left panel of Figure 4). The slower rotating regions at mid–latitudes are also evident at higher distances (from 1.7 to 2.0 $R_\odot$) showed in upper-right and lower-left panel of Figure 4. A careful inspection shows that although the values of the period peaks remain quite constant while getting away from the Sun, their positions tend to move toward lower latitudes with increasing distance. In the mid–latitude regions the rotation periods determined from the available pB data from 1.75 to 2.0 $R_\odot$ are in good agreement with UV results as we show in Figure 4 (upper-right and lower-left panel).

Radial variation

Previous investigations ([18]), using white–light data from K–coronameters, found a nearly constant rotation rate for heights ranging from about 1.1 to 2 $R_\odot$. By using a larger set of data, spanning from the inner to the outer corona, [6] found that the plasma in the inner corona, in particular below 2 $R_\odot$, tends to rotate at a slightly faster rate than the extended corona. In particular, LASCO C1 observations hint at a more or less constant period of 27.4 days much below 2 $R_\odot$ while, between 2.5 and 5 $R_\odot$, LASCO C2 estimates look consistently higher (around 27.65 days) than the LASCO C1 ones. [6] could not constrain the radial gradient of rotation between the two coronal regions due to a gap of several tenths of solar radii in the LASCO C1 and C2 fields of view around 2 $R_\odot$. UVCS observations allow to fill this gap in the equatorial region.

Our result, shown in Figure 4 (lower right panel), obtained by averaging data of three spatial bins around the equator, supports as a real feature the observed discrepancy between the LASCO C1 and C2 rotation periods. The plot indicates that, at least in the equatorial region, the extended corona rotates slower than the inner corona. More precisely, the observed equatorial synodic rotation period is about 0.2 days shorter when the height increases from 1.5 to 2.3 $R_\odot$, while the rotation rate is observed to slow down sensibly with an abrupt transition between about 2.3 and 2.5 $R_\odot$ where the period increases by more than half a day. In general, the average rotation periods in the inner corona (below about 2.3 $R_\odot$) decrease with height at low latitudes (within about ten degrees from the solar equator) and high latitudes (above about 30° from the solar equator) and increase with height about 20° above and below the solar equator.

North-South asymmetry

Asymmetries between the northern and southern hemisphere were previously found in the rotation velocities of photospheric magnetic fields ([5]; [19]). For example, it has been found that the northern hemisphere rotates faster during the even cycles (20 and 22), while the rotation of southern hemisphere dominates in odd ones (cycles 19 and 21) ([20]). A strong asymmetry in the rotation of the north and south hemispheres is also noticeable in Figure 4. In particular, in the northern hemisphere, the rotation looks more solid–body–like (within the errors) and the rotation rate is sensibly slower in the southern hemisphere. This asymmetry might indicate a weak coupling between the magnetic fields of the two hemispheres. Our results also show that, at the start of the cycle 23, the more active hemisphere, the southern one ([21]), rotates somewhat slower.

DISCUSSION AND CONCLUSIONS

We have used synoptic UVCS observations to establish the rotational characteristics of the solar corona at the minimum of solar activity in the time interval from May 1996 to May 1997. The capabilities of UVCS allow the estimate of coronal rotational rates, from a minimum height of 1.5 $R_\odot$ to about 3 $R_\odot$, by studying the recurrence of persistent structures both in UV and visible light.

The coronal rotation rates obtained in this study confirm already established results that the corona, during minimum activity, tends to rotate with a less pronounced differential rotation than the plasma of the photosphere. Moreover, we found a deceleration of the corona towards the photospheric rates at mid–latitudes, where localized flux concentrations of the new–cycle active regions are emerging. This evidence suggests that the corona can be broadly affected by localized flux concentrations (e.g., [**?**]) and that the slower rotation periods can be related to the active regions, that is to the short–lived features, while the faster components can be related to the large–scale global magnetic field.

The study of the radial variation of the coronal rotation period shows that the average rotation periods increase with height at the latitudes that roughly correspond to the transition region between the streamer belt plasma and the coronal hole boundaries at solar minimum (about 20° above and below the solar equator). Moreover in the equatorial region, the rotation, which seems to be more strictly influenced by the coronal large–scale global magnetic field structure, is almost rigid, but the observed equatorial synodic rotation period shows an abrupt increase of about half a day between 2.3 and 2.5 $R_\odot$, where the coronal plasma starts to be dominated by the open field lines extending to the heliosphere. The radial and latitudinal profiles of the coronal rotation seem to track fairly well the expected magnetic topology at solar minimum conditions, in particular, the larger gradients of the rotation rates are localized at the boundary between the open and closed field lines, where the differential photospheric rotation is somehow translated to the corona, as also observed by [7]. The observational evidence suggests possible magnetic shears as a consequence of the rotation profile variation with latitude and distance. In particular, the steep gradient around 2.5 $R_\odot$ can drive a magnetic reconfiguration and thus, a possible energy release to heat the corona, as discussed for example by [6] and [22].

ACKNOWLEDGMENTS

This research has been supported by contract `ASI/I/035/05/0` of the Italian Space Agency. SOHO is a project of international cooperation between ESA and NASA.

REFERENCES

1. Fisher, R., & Sime, D. G., *Astroph. Journal*, **287**, 959 (1984)
2. Antonucci, E., & Svalgaard, L., *Solar Phys.*, **34**, 3 (1974)
3. Aschwanden, M. J., Lim, J., Gary, D. E., & Klimchuk, J. A., *Astroph. Journal*, **454**, 512 (1995)
4. Weber, M. A., Acton, L. W., Alexander, D., Kubo, S., & Hara, H., *Solar Phys.*, **189**, 271 (1999)
5. Antonucci, E., Hoeksema, J. T., & Scherrer, P. H., *Astroph. Journal*, **360**, 296 (1990)
6. Lewis, D. J., et al., *Solar Phys.*, **184**, 297 (1999)
7. Stenborg, G., et al., in Proc. Magnetic Fields and Solar Processes, *ESA SP*-**448**, 1107 (1999)
8. Kohl, J. L., et al., *Solar Phys.* **162**, **313** (1995) (1995)
9. Lomb, N., *Astroph. Journal Suppl. Ser.*, **39**, 447 (1976)
10. Scargle, J. D., *Astroph. Journal*, **263**, 835 (1982)
11. Press, W. H., & Rybicki. G. B., *Astroph. Journal*, **338**, 277 (1989)
12. Horne, J. H., & Baliunas, S. L., *Astroph. Journal*, **302**, 757 (1986)
13. Press, W. H., Numerical recipes in C++ : the art of scientific computing, Cambridge Univ. Press (2002)
14. Parker, G. D., Hansen, R. T., & Hansen, S. F., *Solar Phys.*, **80**, 185 (1982)
15. Sime, D. G., Fisher, R. R., & Altrock, R. C., *Astroph. Journal*, **336**, 454 (1989)
16. Giordano, S., & Mancuso, S.., *Astroph. Journal*, Submitted (2007)
17. Brajša, R., et al., *Solar Phys.*, **206**, 229 (2002)
18. Hansen, R. T., Hansen, S. F., & Loomis, H.G., *Solar Phys.*, **10**, 135 (1969)
19. Javaraiah, J., & Gokhale, M. H., *Solar Phys.*, **170**, 389 (1997)
20. Gigolashvili, M. S., et al., *Solar Phys.*, **227**, 27 (2005)
21. Zharkov, S., Zharkova, V. V., & Ipson, S. S., *Solar Phys.*, **228**, 377 (2005)
22. Simnett, G. M., *Space Science Reviews*, **70**, 69 (1994)

Plasma Flows in Coronal Streamers - Numerical Simulation

Cristiana Dumitrache

Astronomical Institute of the Romanian Academy, Bucharest, Romania, e-mail: crisd@aira.astro.ro

Abstract. Coronal streamers are very interesting, long lived, features of solar corona. They are linked to prominences, CMEs and also to slow solar wind. The have different shape and behavior. We present a numerical simulation of a coronal streamer computed on five solar radii. Vertical motions in the streamer, both ascending and descending, are revealed during the streamer evolution, as well as eddy flows.

Keywords: numerical simulations, corona, coronal streamer, magnetic reconnections
PACS: 96.60.P-, 96.60.pf, 96.60.Iv

INTRODUCTION

Coronal streamers are long lived structures of the solar corona that extend for several solar radii. They are visible in white and UV light. In the last eleven years of SOHO observations our knowledge about these features of solar activity increased considerably. Accurate measurements of plasma parameters and many numerical simulations were performed. The streamers seem to display different morphology in different lines [1] and different characteristics of plasma at different solar radii.

Our study focuses on the flows in coronal streamers. We notice several measurements of velocities reported by different authors: Gopalswamy et al. [2] found velocities of 13 km/s at about 1.19 Rsun, Noci et al. [3] showed that rapid velocities occur on the lateral side of the streamer and there are higher when the distance from the symmetry axis of the streamer is bigger. Radial outflow velocities were determined from Doppler dimming observations from SOHO/UVCS by Kohl [4]. They measured plasma flows for 1.25 to 10 Rsun and found for H^0 decreasing outflows from 180 km/s at 2 Rsun to about 140 km/s at 8 Rsun. The values for O^{5+} increase and reach the maximum of 150 km/s at 4.7 Rsun.

Wang [5] reported unexpected small-scale phenomena, including plasma blobs that are ejected continually from the cusplike bases of streamers and inflows that occur mainly during times of high solar activity. These phenomena were revealed in white light images acquired from space.

The observations suggest that open and closed field lines reconnect near the cusp of the streamer to form blobs of higher plasma density that are ejected into the slow solar wind. Lapenta [6] simulated flows in coronal streamers and conclude that reconnections produced by converging flows are responsible for the plasma blobs ejections.

CP934, *Flows, Boundaries, Interactions*
edited by C. Dumitrache, V. Mioc, and N. A. Popescu

NUMERICAL SIMULATION

We performed a numerical experiment on 10 solar radii, starting with a current sheet initial configuration and reflecting outer boundary conditions. Initial values of plasma parameters are: $\beta = 0.5$, $Rm = 10^{-3}$, $\sigma = 10^{-1}$. The figures below describe the evolution of the magnetic field topology, in the left panels, and the flows (on half of grid) superposed on density contour plots (filled, entire the grid), in the right panels.

At the Alfvén time $t = 0.492$ (fig.1) we obtained a current sheet that has the tendency to huddle. The coronal matter is pushed from the lateral side of the streamer inside it and plasma is collected at the stream base, in many stratified zones. At $t = 2.729$ (fig.2) magnetic reconnections occurs and the field lines close under $5-6$ Rsun: the streamer like magnetic configuration is obtained at this time. The flows are still oriented from lateral side, plasma being pushed in the sheet.

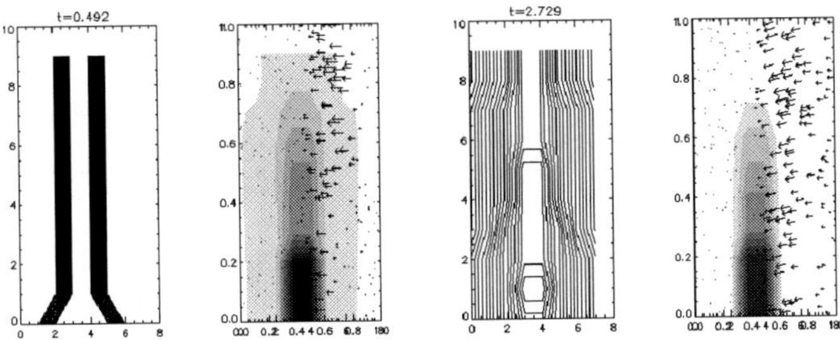

FIGURE 1. Results at t=0.492 and t=2.729

At $t = 34.060$ (fig.2) the magnetic configuration display closed field lines on the central part and along the streamer. Closed islands could be observed and plasma flow comes also from lateral side of the streamer, sloped and changing of the orientation at the streamer wall: ascendent flows outside the streamer and descendent inside it. The low conductivity permits plasma to transverse the magnetic field lines and the lateral wall changes the flow direction at this moment. At $t = 34.189$ plasma is pushed down and the helmet flattened. The flows become descendent and divergent and plasma density stratification disappears for the moment: it rests only the core of the streamer with a

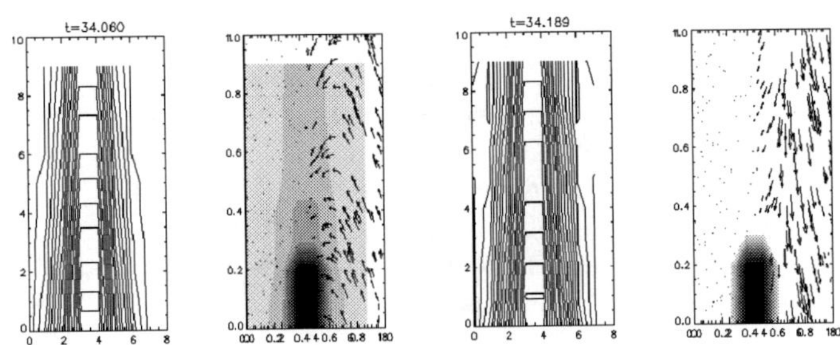

FIGURE 2. Results at t=34.060 and t=34.189

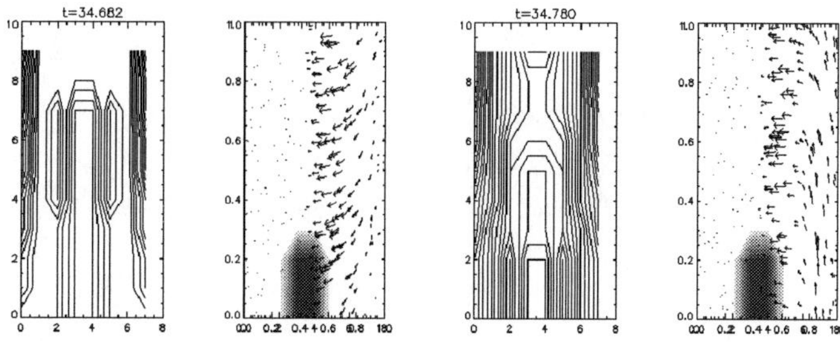

FIGURE 3. Results at t=34.682 and t=34.780

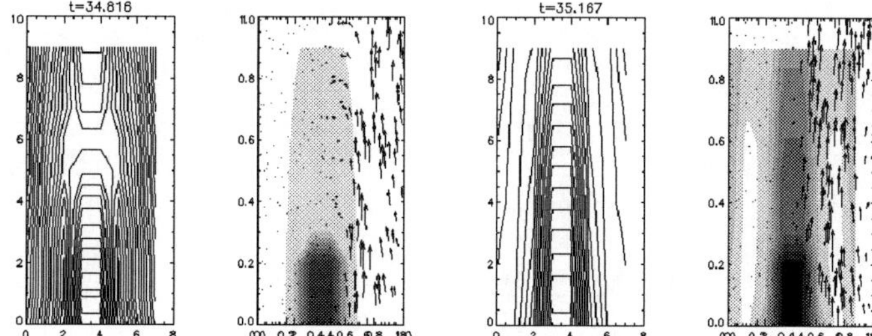

FIGURE 4. Results at t=34.816 and t=35.167

height of about 3 Rsun.

Magnetic reconnections occur at $t = 34.682$ (fig.3) when the flows change again the direction and become slowly lateral ones. New magnetic reconnections produce at $t = 34.780$, when flows change again the direction, from lateral-downward to lateral upward. The magnetic field lines display a configuration (fig. 3, right) that suggest ejected island leaving the Sun. This behavior is more evident at $t = 34.816$ (fig.4), when the magnetic structure is cuttled off inside the streamer at about 6 Rsun. The flows are ascendent and the helmet of the streamer reforms.

A well formed helmet streamer structure is obtained at $t = 35.167$ (fig.4): the density contour plots display a stratification and the cusp point is observed at about 3 Rsun. Plasma flow is only outward until $t = 35.513$ (fig.5), when the magnetic field topology indicates gravitational instabilities. Flows change the direction becoming downward and the density contour plots suggest that plasma was cuttled of and insulated. Probably, magnetic blobs were ejected previously as a consequence of magnetic reconnections and the dramatically changes of directions of the plasma flows.

At $t = 35.878$ (fig.6) plasma movement is upward oriented and it is divergent: part of flows are laterally oriented and to outside the streamer. At this Alfvén time, corresponding to 20.763 days from the beginning of the structure evolution, tearing modes appear. The helmet is also stratified - we observe a cusp point at 3 Rsun and another important point, as a secondary cusp point, at about 6 Rsun.

Another dramatically change of flows direction appears at $t = 35.988$ (fig.7), when

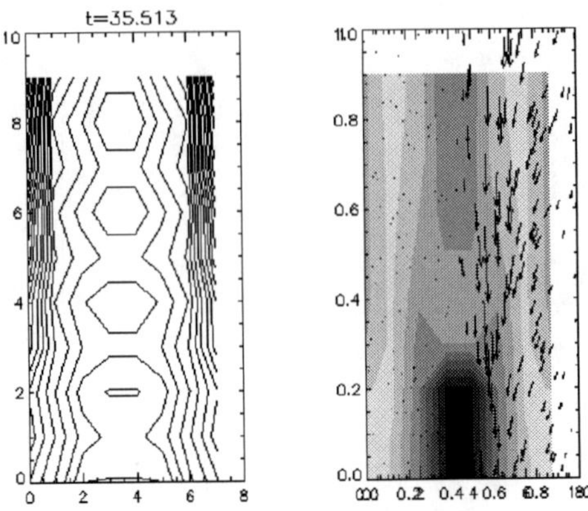

FIGURE 5. Results at t=35.513

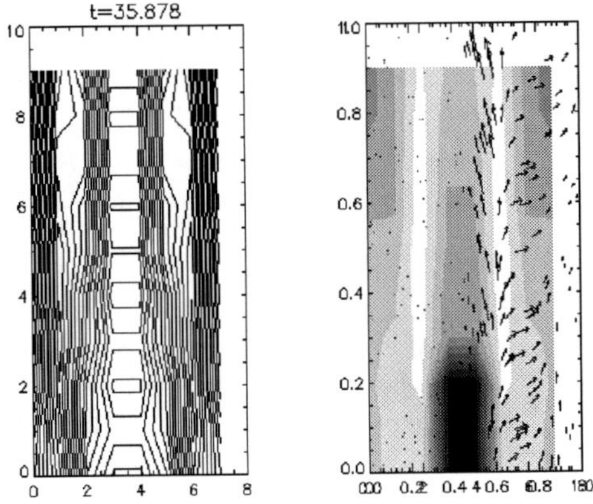

FIGURE 6. Results at t=35.878

plasma inflows downward from lateral side of the streamer and turns in upward flows inside the streamer. Magnetic reconnections occur and magnetic island are ejected from the streamer - not only blobs from the current sheet are ejected, but also parts of the lateral wall of the streamer reconnect and are expelled.

The scenario repeats many times: changes of the plasma flows alternating in downward, lateral inflows and upward directions, having as consequence magnetic reconnections and magnetic blobs ejections. The flows are either convergent, or divergent. Especially, the downward flows are divergent and produce gravitational instabilities. Important points in the changes of flows orientation are the primary and secondary cusp points. In figure 8 (right panel) we observe at the Alfvén time $t = 36.244$ downward divergent flows that have an inflexion point for the inflows at about 8 Rsun where the direction changes and we see upward flows (eddy flows). The separation between the

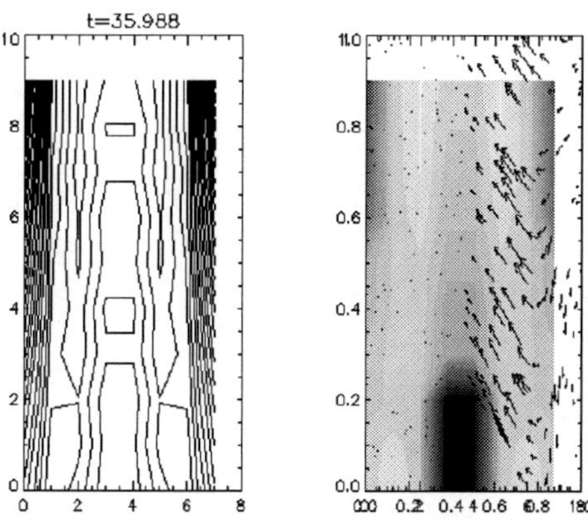

FIGURE 7. Results at t=35.988

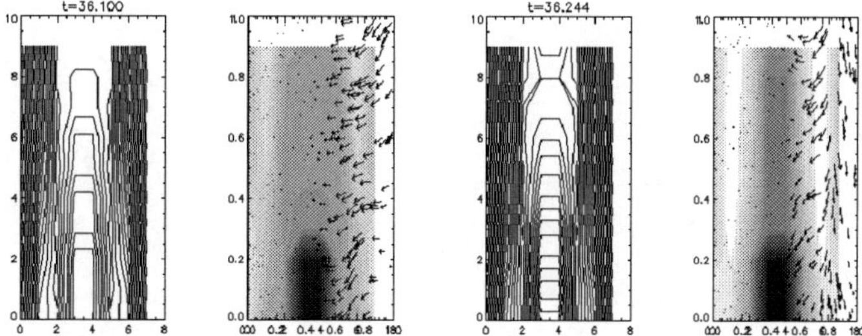

FIGURE 8. Results at t=36.100 and t=36.244

regions above and below 8 Rsun could be seen also on the magnetic topology.

From this moment ($t = 36.244$) we consider two cases in the streamer evolution: (a) the follow up of the simulation until the streamer evanescence dissipating into the solar corona; (b) a mass injection which drives a CME.

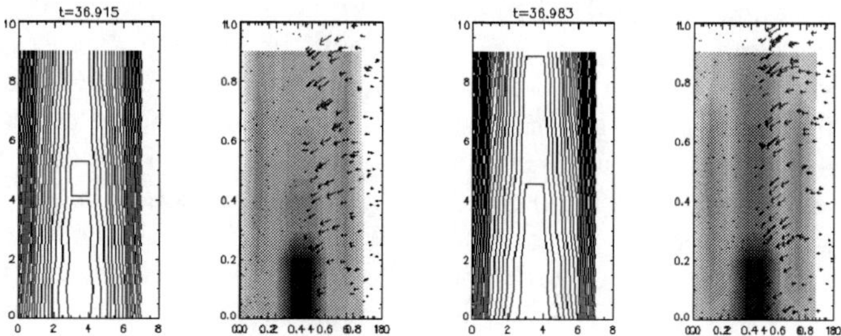

FIGURE 9. Results at t=36.915 and t=36.983

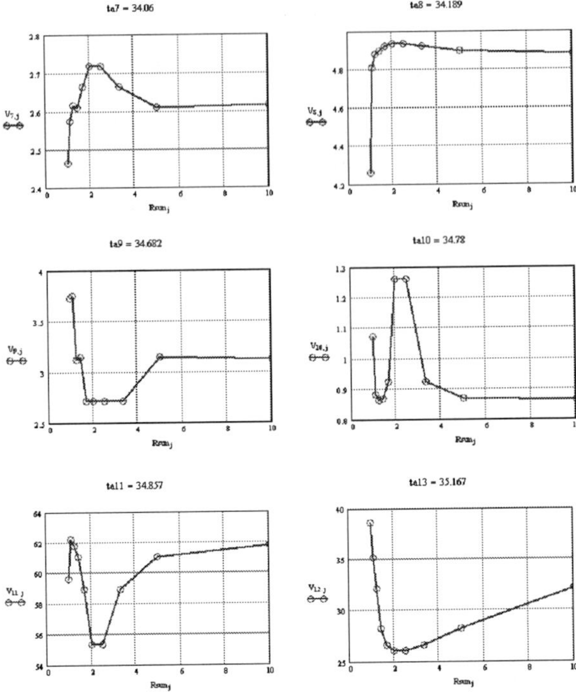

FIGURE 10. Velocity values plots inside the streamer

The magnetic field configuration, contour plots of the density and plasma flows of streamer evolution are displayed in figures 9. The streamer is pinched and restrained by down-flows. The streamer dissipates after more than 21 days from its formation.

Figures 10 and 11 display the time evolution of the velocity values, at different solar radii and at different moments. At the beginning when the velocities have small values, it can be observed an acceleration at about 2 Rsun and a deceleration at about 2.5 – 3 Rsun. Then the velocity changes its behavior at Alfvén time $t = 34.78$ and then at $t = 34.857$, having an increase of the values from undetectable to observable ones. This happened again at $t = 36.1$ and $t = 36.24$, when the values increase more, as displayed in figures 10 and 11. We notice that the importance of the point situated at 5 Rsun could be seen also in figures 10 and 11 - below this point there are significant acceleration and deceleration of flows. In the last stages of the streamer life, the plasma flow decelerates after 1.5 Rsun and accelerates after 3.3 Rsun. The velocity attaints about 60 km/s after 5 solar radii.

In the case (b) we considered a chromospheric mass injection inside the streamer on a narrow region at its symmetry axis, starting with $t = 36.244$ (fig.8, right). Fast mode magnetic reconnections occur at $t = 36.253$ and almost instantaneous a coronal mass ejection onset is registered. This ejection attaints 1200 km/s and the main body of the ejected blob has a length of about 3 solar radii (figure 13). This kind of behavior could produce in the active regions. Figure 14 displays the CME velocity vs Rsun and also the

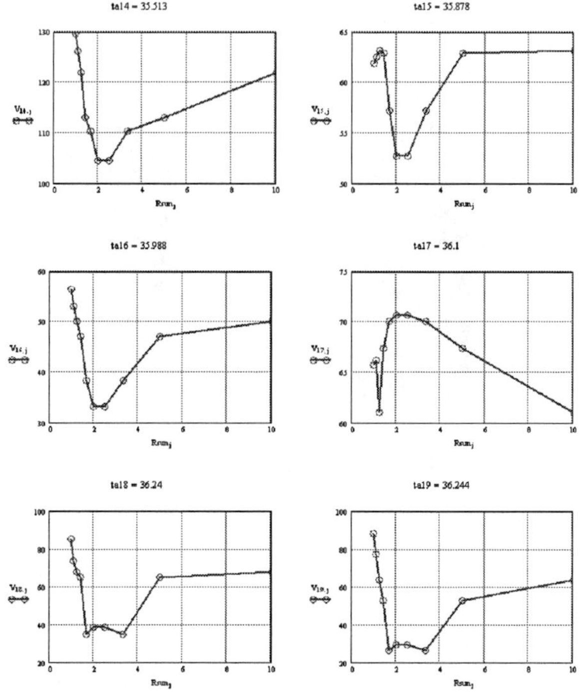

FIGURE 11. Velocity values plots and behavior inside the streamer

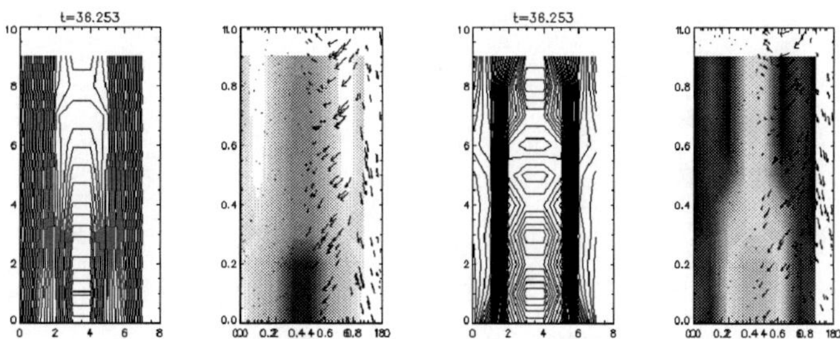

FIGURE 12. Results at t=36.253

density values vs. time.

CONCLUSIONS

We have performed a numerical simulation on 10 Rsun starting with a current sheet initial configuration. The current sheets form naturally in the solar atmosphere. A helmet

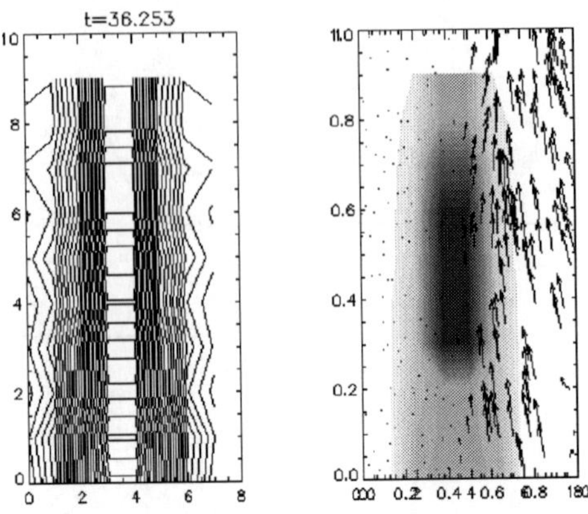

FIGURE 13. Results at t=36.253 -CME produces

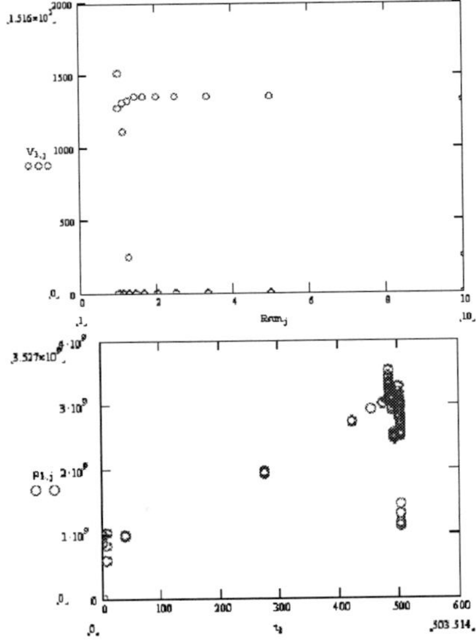

FIGURE 14. Velocity and density evolution in the case of CME

streamer forms after 1.5 days and evolves almost 21 days. The scenarios of its evolution are (a) streamer that disappears quiet in the solar atmosphere and (b) streamer that erupts in a big CME, as a consequence of a mass injection from below.

We have studied especially the plasma flows inside the streamer: the initially lateral flows change the direction becoming periodically downward and upward. Tearing modes and gravitational instabilities (at downward flows) occur. The velocity values increase in time at the distances greater than 5 solar radii. The region contended between 1.5 and

3.3 Rsun is very important in deceleration and than the acceleration of plasma.

Several points are important in the flows regime: the changing in direction occurs preferentially near the cusp point of the streamer, at 3 solar radii and at 6 solar radii. The flows are convergent but frequently also divergent (especially the down flows). Periodically magnetic reconnections produce and plasma blobs are ejected, with low velocity values (that could not be observed) at the beginning stages of streamer life. Eddy flows cloud appear above 8 Rsun.

This numerical experiment reveals the dynamics of the matter in a coronal streamer. These kinds of flows were observed by many authors, but we have obtained also velocity values under the observational limits - this does not mean that they do not exist.

REFERENCES

1. G.Noci, J.L.Kohl, E.Antonucci, et.al., in Proc.V-th SOHO Workshop, *ESA-SP* **404**, 75-84, 1997
2. N. Gopalswamy, M. Shimojo, W. Lu, S. Yashiro, K. Shibasaki, R.A. Howard, *Adv. Space Res.***33**, 676-680, 2004
3. G.Noci, E.Gavriuseva, *ApJ* **658**, L63-L66 , 2007
4. J.L.Kohl, G.Noci, E.Antonucci, G.Tondello, M.C.E.Huber, L.D.Gardner, P.Nicolosi, et al., *SolPhys* **175**, 613-644, 1997
5. Y.-M.Wang, N. R.Sheeley, D. G.Socker, R. A.Howard, N. B.Rich, *J. Geophys. Res.***105**, Issue A11, 25133-25142, 2000
6. G.Lapenta, D.A.Knoll, *ApJ* **624**, 1049-1056, 2005

Heliospheric Electric and Magnetic Fields

Adrian Sabin Popescu

*Astronomical Institute of Romanian Academy,
Str. Cutitul de Argint 5, RO-040557 Bucharest, Romania*

Abstract. From the Maxwell equations in the local Minkowski spacetime chart (derived from the DEUS topology) we obtain the relations to be particularized for a solar type star and a massive star, and later to be used for a 3D representation of the electric and magnetic field topology (in heliosphere or in a stellar atmosphere) and of its evolution with the cosmological time.

Keywords: DEUS topology, Maxwell equations, heliosphere, stellar atmospheres
PACS: 96.60.Hv , 96.60.Vg , 97.10.Ex , 97.10.Ld

INTRODUCTION

This paper is having a modest goal: to obtain the equation background of a code meant to follow the evolution of the electric and magnetic field topology in heliosphere (or solar type star atmosphere) and in the atmosphere of a massive star (or black hole). The difference between the two cases will come from the choice of the local Minkowski coordinate chart [1].

MAXWELL EQUATIONS

In Minkowski spacetime, in curvilinear coordinates:

$$\nabla\times\mathbf{E} = \mp\frac{2}{\sqrt{3}}\,\frac{1}{r}\left[\frac{1}{\tan\zeta}\,E_\xi + \frac{\partial E_\xi}{\partial\zeta} - \frac{1}{\sin\zeta}\,\frac{\partial E_\zeta}{\partial\xi}\right]\vec{e_r} \mp$$
$$\mp\frac{2}{\sqrt{3}}\,\frac{1}{r}\left[\frac{1}{\sin\zeta}\,\frac{\partial E_r}{\partial\xi} - E_\xi - r\,\frac{\partial E_\xi}{\partial r}\right]\vec{e_\zeta} \mp \frac{2}{\sqrt{3}}\,\frac{1}{r}\left[E_\zeta + r\,\frac{\partial E_\zeta}{\partial r} - \frac{\partial E_r}{\partial\zeta}\right]\vec{e_\xi}\,. \tag{1}$$

We have also:

$$\frac{\partial\mathbf{B}}{\partial t} = \frac{\partial t_{FRW}}{\partial t}\left[\frac{\partial B_r}{\partial t_{FRW}}\,\vec{e_r} + \frac{\partial B_\zeta}{\partial t_{FRW}}\,\vec{e_\zeta} + \frac{\partial B_\xi}{\partial t_{FRW}}\,\vec{e_\xi}\right] =$$
$$= \pm\frac{1}{\cosh\left[\arctan\left(\frac{1}{\tan t_{FRW}}\right)\right]}\left[\frac{\partial B_r}{\partial t_{FRW}}\,\vec{e_r} + \frac{\partial B_\zeta}{\partial t_{FRW}}\,\vec{e_\zeta} + \frac{\partial B_\xi}{\partial t_{FRW}}\,\vec{e_\xi}\right]. \tag{2}$$

By combining (1) and (2) in the Faraday's law of induction results:

CP934, *Flows, Boundaries, Interactions*
edited by C. Dumitrache, V. Mioc, and N. A. Popescu

$$\frac{2}{\sqrt{3}}\frac{1}{r}\left[\frac{1}{\tan\zeta}E_\xi+\frac{\partial E_\xi}{\partial\zeta}-\frac{1}{\sin\zeta}\frac{\partial E_\zeta}{\partial\xi}\right]=\frac{1}{\cosh\left[\arctan\left(\frac{1}{\tan t_{FRW}}\right)\right]}\frac{\partial B_r}{\partial t_{FRW}}, \tag{3}$$

$$\frac{2}{\sqrt{3}}\frac{1}{r}\left[\frac{1}{\sin\zeta}\frac{\partial E_r}{\partial\xi}-E_\xi-r\frac{\partial E_\xi}{\partial r}\right]=\frac{1}{\cosh\left[\arctan\left(\frac{1}{\tan t_{FRW}}\right)\right]}\frac{\partial B_\zeta}{\partial t_{FRW}}, \tag{4}$$

$$\frac{2}{\sqrt{3}}\frac{1}{r}\left[E_\zeta+r\frac{\partial E_\zeta}{\partial r}-\frac{\partial E_r}{\partial\zeta}\right]=\frac{1}{\cosh\left[\arctan\left(\frac{1}{\tan t_{FRW}}\right)\right]}\frac{\partial B_\xi}{\partial t_{FRW}}. \tag{5}$$

In the same manner, we can write the Gauss law of magnetism:

$$\nabla.\mathbf{B}=\pm\frac{2}{\sqrt{3}}\frac{1}{r}\left[r\frac{\partial B_r}{\partial r}\vec{e_r}+\left(\frac{1}{\tan\zeta}B_\zeta+\frac{\partial B_\zeta}{\partial\zeta}\right)\vec{e_\zeta}+\frac{1}{\sin\zeta}\frac{\partial B_\xi}{\partial\xi}\vec{e_\xi}\right]=0, \tag{6}$$

resulting:

$$\frac{\partial B_r}{\partial r}=0, \tag{7}$$

$$\frac{1}{\tan\zeta}B_\zeta+\frac{\partial B_\zeta}{\partial\zeta}=0, \tag{8}$$

$$\frac{\partial B_\xi}{\partial\xi}=0. \tag{9}$$

This means that:

$$\frac{\partial B_r}{\partial t_{FRW}}=\frac{\partial B_r}{\partial r}\frac{\partial r}{\partial t_{FRW}}=0, \tag{10}$$

which gives in (3):

$$\frac{1}{\tan\zeta}E_\xi+\frac{\partial E_\xi}{\partial\zeta}-\frac{1}{\sin\zeta}\frac{\partial E_\zeta}{\partial\xi}=0, \tag{11}$$

and:

$$\frac{\partial B_\xi}{\partial t_{FRW}}=\frac{\partial B_\xi}{\partial\xi}\frac{\partial\xi}{\partial t_{FRW}}=0, \tag{12}$$

which gives in (5):

$$E_\zeta+r\frac{\partial E_\zeta}{\partial r}-\frac{\partial E_r}{\partial\zeta}=0. \tag{13}$$

Also, with:

$$\frac{\partial B_\zeta}{\partial t_{FRW}} = \frac{\partial B_\zeta}{\partial \zeta}\frac{\partial \zeta}{\partial t_{FRW}} = -\frac{1}{\tan\zeta} B_\zeta \,, \tag{14}$$

where, from [1], $\frac{\partial \zeta}{\partial t_{FRW}} = 1$, in (4), we get:

$$\frac{2}{\sqrt{3}}\frac{1}{r}\left[\frac{1}{\sin\zeta}\frac{\partial E_r}{\partial \xi} - E_\xi - r\frac{\partial E_\xi}{\partial r}\right] = -\frac{1}{\cosh\left[\arctan\left(\frac{1}{\tan t_{FRW}}\right)\right]}\frac{1}{\tan\zeta} B_\zeta \,. \tag{15}$$

When we consider the [1] coordinates for a solar type star, as function of t_{FRW}, the relation (11) can be written as:

$$-\tan t_{FRW}\, E_\xi + \frac{\partial E_\xi}{\partial t_{FRW}} + \frac{1}{\sin\left[\arctan\left(\frac{1}{\tan t_{FRW}}\right)\right]}\frac{1}{c_3\cosh\left[\arctan\left(\frac{1}{\tan t_{FRW}}\right)\right]}\frac{\partial E_\zeta}{\partial t_{FRW}} = 0 \,, \tag{16}$$

relation (13) can be written as:

$$E_\zeta + \frac{1}{c_1\cosh\left[\arctan\left(\frac{1}{\tan t_{FRW}}\right)\right]}\frac{\partial E_\zeta}{\partial t_{FRW}} - \frac{\partial E_r}{\partial t_{FRW}} = 0 \,, \tag{17}$$

and relation (15) as:

$$-\frac{2}{\sqrt{3}}\frac{1}{c_2\exp\left\{-c_1\sinh\left[\arctan\left(\frac{1}{\tan t_{FRW}}\right)\right]\right\}}\left\{\frac{1}{c_3\sin\left[\arctan\left(\frac{1}{\tan t_{FRW}}\right)\right]}\frac{\partial E_r}{\partial t_{FRW}} + \right.$$
$$\left. + \cosh\left[\arctan\left(\frac{1}{\tan t_{FRW}}\right)\right] E_\xi + \frac{1}{c_1}\frac{\partial E_\xi}{\partial t_{FRW}}\right\} = \tan t_{FRW}\, B_\zeta \,. \tag{18}$$

Now, in natural units, in Ampère's law we have $\mu_0\epsilon_0 = 1/c^2 = 1$. The electric current density will be described by $\mathbf{J} = \rho\,\mathbf{v}$ or, with the help of Gauss law for electricity:

$$\mathbf{J} = \epsilon_0\,\mathbf{v}(\nabla.\mathbf{E}) \,. \tag{19}$$

In (19), the velocity $\mathbf{v} = dx^i/dt$, with $x^i = \{r,\zeta,\xi\}$ and t the global DEUS time:

$$\frac{dt}{dt_{FRW}} = \pm\cosh\left[\arctan\left(\frac{1}{\tan t_{FRW}}\right)\right] . \tag{20}$$

Then, we can write the Ampère's law as:

$$\nabla\times\mathbf{B} = \mathbf{v}(\nabla.\mathbf{E}) + \frac{\partial \mathbf{E}}{\partial t} \,. \tag{21}$$

In our curvilinear Minkowski coordinates:

$$\nabla\times\mathbf{B} = \mp\frac{2}{\sqrt{3}}\,\frac{1}{r}\left[\frac{1}{\tan\zeta}\,B_\xi + \frac{\partial B_\xi}{\partial\zeta} - \frac{1}{\sin\zeta}\,\frac{\partial B_\zeta}{\partial\xi}\right]\vec{e_r} \mp \\ \mp\frac{2}{\sqrt{3}}\,\frac{1}{r}\left[\frac{1}{\sin\zeta}\,\frac{\partial B_r}{\partial\xi} - B_\xi - r\,\frac{\partial B_\xi}{\partial r}\right]\vec{e_\zeta} \mp \frac{2}{\sqrt{3}}\,\frac{1}{r}\left[B_\zeta + r\,\frac{\partial B_\zeta}{\partial r} - \frac{\partial B_r}{\partial\zeta}\right]\vec{e_\xi}, \tag{22}$$

or, using (7) and (9) results of the Gauss law for magnetism:

$$\nabla\times\mathbf{B} = \mp\frac{2}{\sqrt{3}}\,\frac{1}{r}\left[\frac{1}{\tan\zeta}\,B_\xi - \frac{1}{\sin\zeta}\,\frac{\partial B_\zeta}{\partial\xi}\right]\vec{e_r} \mp \frac{2}{\sqrt{3}}\,\frac{1}{r}\left[-B_\xi\right]\vec{e_\zeta} \mp \frac{2}{\sqrt{3}}\,\frac{1}{r}\left[B_\zeta + r\,\frac{\partial B_\zeta}{\partial r}\right]\vec{e_\xi}. \tag{23}$$

We can explicitly write also:

$$\mathbf{v}(\nabla.\mathbf{E}) = \pm\frac{2}{\sqrt{3}}\,\frac{\partial t_{FRW}}{\partial t}\left[\frac{\partial E_r}{\partial t_{FRW}}\,\vec{e_r} - \frac{\tan t_{FRW}}{r}\,E_\zeta\,\vec{e_\zeta} + \frac{1}{r}\,\frac{\partial E_\zeta}{\partial t_{FRW}}\,\vec{e_\zeta} + \frac{1}{r\,\sin\zeta}\,\frac{\partial E_\xi}{\partial t_{FRW}}\,\vec{e_\xi}\right]. \tag{24}$$

Then, from the (21) law, with (8), we obtain:

$$\frac{\partial E_r}{\partial t_{FRW}} = \mp\frac{2}{\sqrt{3}}\,\frac{\partial t}{\partial t_{FRW}}\,\frac{1}{1\pm\frac{2}{\sqrt{3}}}\,\frac{1}{r}\left[\frac{1}{\tan\zeta}\,B_\xi + \frac{1}{\sin\zeta}\,\frac{1}{\tan\zeta}\,B_\zeta\,\frac{\partial t_{FRW}}{\partial\xi}\right], \tag{25}$$

$$B_\xi = \frac{\partial t_{FRW}}{\partial t}\left[-\tan t_{FRW}\,E_\zeta + \frac{\partial E_\zeta}{\partial t_{FRW}}\right] \pm \frac{\sqrt{3}}{2}\,r\,\frac{\partial E_\zeta}{\partial t_{FRW}}\,\frac{\partial t_{FRW}}{\partial t}, \tag{26}$$

$$\frac{\partial E_\xi}{\partial t_{FRW}} = \mp\frac{2}{\sqrt{3}}\,\frac{\partial t}{\partial t_{FRW}}\,\frac{\sin\zeta}{r\,\sin\zeta\pm\frac{2}{\sqrt{3}}}\left[1 - \frac{r}{\tan\zeta}\,\frac{\partial t_{FRW}}{\partial r}\right]B_\zeta. \tag{27}$$

With (26) in (25) we get:

$$\frac{\partial E_r}{\partial t_{FRW}} = \mp\frac{2}{\sqrt{3}}\,\frac{1}{1\pm\frac{2}{\sqrt{3}}}\,\frac{1}{r}\left\{\frac{1}{\tan\zeta}\left[-\tan t_{FRW}\,E_\zeta + \frac{\partial E_\zeta}{\partial t_{FRW}}\right]\pm \\ \pm\frac{\sqrt{3}}{2}\,\frac{r}{\tan\zeta}\,\frac{\partial E_\zeta}{\partial t_{FRW}} + \frac{1}{\sin\zeta}\,\frac{1}{\tan\zeta}\,B_\zeta\,\frac{\partial t_{FRW}}{\partial\xi}\,\frac{\partial t}{\partial t_{FRW}}\right\}. \tag{28}$$

Expressing equation (13) with the help of (28), we have:

$$E_\zeta + r\,\frac{\partial E_\zeta}{\partial t_{FRW}}\,\frac{\partial t_{FRW}}{\partial r} \pm \frac{2}{\sqrt{3}}\,\frac{1}{1\pm\frac{2}{\sqrt{3}}}\,\frac{1}{r}\left\{\frac{1}{\tan\zeta}\left[-\tan t_{FRW}\,E_\zeta + \frac{\partial E_\zeta}{\partial t_{FRW}}\right]\pm \\ \pm\frac{\sqrt{3}}{2}\,\frac{r}{\tan\zeta}\,\frac{\partial E_\zeta}{\partial t_{FRW}} + \frac{1}{\sin\zeta}\,\frac{1}{\tan\zeta}\,B_\zeta\,\frac{\partial t_{FRW}}{\partial\xi}\,\frac{\partial t}{\partial t_{FRW}}\right\} = 0. \tag{29}$$

Equation (11) with (27) becomes:

$$\frac{1}{\tan\zeta}E_\xi \mp \frac{2}{\sqrt{3}}\frac{\partial t}{\partial t_{FRW}}\frac{\sin\zeta}{r\sin\zeta \pm \frac{2}{\sqrt{3}}}\left[1-\frac{r}{\tan\zeta}\frac{\partial t_{FRW}}{\partial r}\right]B_\zeta - \frac{1}{\sin\zeta}\frac{\partial E_\zeta}{\partial t_{FRW}}\frac{\partial t_{FRW}}{\partial \xi} = 0\,. \tag{30}$$

From equation (15) results:

$$B_\zeta = -\frac{2}{\sqrt{3}}\frac{1}{r}\cosh\left[\arctan\left(\frac{1}{\tan t_{FRW}}\right)\right]\tan\zeta\left[\frac{1}{\sin\zeta}\frac{\partial E_r}{\partial t_{FRW}}\frac{\partial t_{FRW}}{\partial \xi} - E_\xi - r\frac{\partial E_\xi}{\partial t_{FRW}}\frac{\partial t_{FRW}}{\partial r}\right], \tag{31}$$

and then, with (27) and (28):

$$\begin{aligned} B_\zeta = &-\frac{2}{\sqrt{3}}\frac{1}{r}\cosh\left[\arctan\left(\frac{1}{\tan t_{FRW}}\right)\right]\tan\zeta\left\{1\mp\frac{2}{\sqrt{3}}\frac{1}{r}\cosh\left[\arctan\left(\frac{1}{\tan t_{FRW}}\right)\right]\tan\zeta\,\times\right.\\ &\times\left[\frac{1}{\sin^2\zeta}\frac{2}{\sqrt{3}}\frac{1}{1\pm\frac{2}{\sqrt{3}}}\frac{1}{r}\frac{1}{\tan\zeta}\left(\frac{\partial t_{FRW}}{\partial \xi}\right)^2\frac{\partial t}{\partial t_{FRW}} - r\frac{2}{\sqrt{3}}\frac{\partial t}{\partial t_{FRW}}\frac{\sin\zeta}{r\sin\zeta\pm\frac{2}{\sqrt{3}}}\,\times\right.\\ &\left.\left.\times\left(1-\frac{r}{\tan\zeta}\frac{\partial t_{FRW}}{\partial r}\right)\frac{\partial t_{FRW}}{\partial r}\right]\right\}^{-1}\left\{\mp\frac{1}{\sin\zeta}\frac{2}{\sqrt{3}}\frac{1}{1\pm\frac{2}{\sqrt{3}}}\frac{1}{r}\,\times\right.\\ &\left.\times\left[-\frac{\tan t_{FRW}}{\tan\zeta}E_\zeta + \frac{1}{\tan\zeta}\frac{\partial E_\zeta}{\partial t_{FRW}} \pm \frac{\sqrt{3}}{2}\frac{r}{\tan\zeta}\frac{\partial E_\zeta}{\partial t_{FRW}}\right]\frac{\partial t_{FRW}}{\partial \xi} - E_\xi\right\}. \end{aligned} \tag{32}$$

CONCLUSIONS

Using the (r,ζ,ξ) coordinates for the solar type star or for a massive star (or black hole) [1] we can study how the electric and magnetic field components vary according to the t_{FRW} cosmic time.

With (32) in (29) and (30), we are able to determine E_ζ and E_ξ, from where, back in (32), B_ζ and, in (28), E_r. As resulting from the Gauss law for magnetism, B_ξ and B_r are constant through the t_{FRW} cosmologic time.

We checked if the system formed with (29) and (30) equations has solution at $t_{FRW} \simeq 0.524$ and found that it does, even that it is very complicated and dependent on the constants c_1 - c_3. A fine tuning of these constants will be possible by comparing the result of the 3D simulation with the observed heliospheric electric and magnetic field topology.

With a variation in distance from the solar surface to the heliosphere limit ($c_2 \in (0,70]$), at a constant time (e.g, $t_{FRW} = 0.524$), we can get a snapshot of the electric and magnetic field topology into the heliosphere. For this, as initial value for the fields we take $B_{0r} = B_{0\zeta} = B_{0\xi} = 0$ and $E_{0r} = E_{0\zeta} = E_{0\xi}$, obtained at $t_{FRW} = 0.5$ from:

$$\frac{1}{\cosh\left[\arctan\left(\frac{1}{\tan t_{FRW}}\right)\right]} \simeq 10^{-38} \to 0$$

and

$$\exp\left\{-c_1 \sinh\left[\arctan\left(\frac{1}{\tan t_{FRW}}\right)\right]\right\} \simeq \exp\left(-10^{38}\right) \to 0$$

REFERENCES

1. A.S. Popescu, "Gravito-magnetic Heliospheric Surfaces", *AIP Conference Proceedings Series: Fifty Years of Romanian Astrophysics*, Eds. C. Dumitrache et al., ISBN 978-0-7354-0400-7 (2007).

Fractional diffusion equations and applications

Emil Popescu

Technical University of Civil Engineering, Bucharest, Romania, e-mail:epopescu@utcb.ro

Abstract. This paper presents a method for an explicit analysis of the equations with fractional derivatives that describe important physical processes in solar wind plasmas, in plasmas of thermonuclear devices, etc. Space-time fractional diffusions account for anomalous features, which are observed in such physical processes. In certain cases the fundamental solutions of these equations can be interpreted as probability density functions. Thus, we observe that anomalous diffusion equations are related to Levy stable non-Gaussian processes. An example is the multiscale nature of the magnetosphere, where the correlated data of the solar wind-magnetosphere system show that probability distribution function is non-Gaussian.

Keywords: anomalous diffusion, multiscale behavior, coupled space-time fractional derivative equation
PACS: 02.30.Jr, 02.50.-r, 96.25.Qr

FRACTIONAL DERIVATIVES AND ANOMALOUS DIFFUSIONS

Differentiation is usually regarded as discrete operation, in the sense that we differentiate a function once, twice, or any whole number of times. However, in some circumstances it is useful to evaluate a fractional derivative.

Fractional derivatives $\frac{d^\alpha f}{dx^\alpha}(x)$ for any $\alpha > 0$ were invented by Leibnitz soon after the more familiar integer derivatives.

One extends to the fractional case some derivative formulas:

$$\frac{d^\alpha f}{dx^\alpha}\left(e^{\lambda x}\right) = \lambda^\alpha e^{\lambda x}, \ \frac{d^\alpha f}{dx^\alpha}(x^r) = \frac{\Gamma(r+1)}{\Gamma(r-\alpha+1)} x^{r-\alpha}, \tag{1}$$

where $\Gamma(r) = \int_0^\infty x^{r-1} e^{-x} dx$ for $r > 0$, $\Gamma(p+1) = p!$ if $p \in \mathbf{N}$.

If the Fourier transform of $f(x)$ is defined for $k \in \mathbf{R}$ by

$$\widehat{f}(k) = \int_{-\infty}^{\infty} e^{-ikx} f(x)\, dx, \tag{2}$$

then $\frac{d^\alpha f}{dx^\alpha}(x)$ has Fourier transform $(ik)^\alpha \widehat{f}(k)$, where $(ik)^\alpha = |k|^\alpha e^{isign(k)\alpha\pi/2}$.

To solve the fractional diffusion equation

$$\frac{\partial u}{\partial t}(x,t) = \frac{\partial^\alpha u}{\partial x^\alpha}(x,t) \tag{3}$$

for $0 < \alpha \leq 2$ we apply the Fourier transform and we obtain

$$\frac{d\widehat{u}}{dt}(k,t) = (ik)^\alpha \widehat{u}(k,t). \tag{4}$$

CP934, *Flows, Boundaries, Interactions*
edited by C. Dumitrache, V. Mioc, and N. A. Popescu

If we take the initial condition $\widehat{u}(k,t)\mid_{t=0}=1$, we get $\widehat{u}(k,t)=e^{t(ik)^{\alpha}}$. Inverting this Fourier transform, we have the solution of the fractional diffusion equation which is called $\alpha-$stable density or Levy distribution for every $t>0$.

If $\alpha=2$, then this solution is the Gaussian probability density

$$u(x,t)=\frac{1}{\sqrt{4\pi t}}e^{-x^2/4t},\quad t>0. \tag{5}$$

$u(x,t)$ denotes the relative concentration of particles at location x at time t. The stable distributions are the only distributions that can be obtained as limits of normalized sums of independent identically distributed random variables.

Let X be a random variable, which has the density $f(x)$, such that

$$P(a\leq X\leq b)=\int_a^b f(x)\,dx. \tag{6}$$

Then

$$\begin{aligned}\widehat{f}(k)=\textstyle\int_{-\infty}^{\infty}e^{-ikx}f(x)\,dx=\int_{-\infty}^{\infty}\left(1-ikx+\frac{1}{2!}(ikx)^2-\cdots\right)f(x)\,dx=\\ =1-ik\mu_1-\tfrac{1}{2}k^2\mu_2+\cdots,\end{aligned} \tag{7}$$

where $\mu_p=\int_{-\infty}^{\infty}x^p f(x)\,dx$ is the $p-$th moment of X.

If $\mu_1=0$ and $\mu_2=2$ then we obtain a "nice" formula: $\widehat{f}(k)=1-k^2+\cdots$.

If X_n represents a particle jump at time n then $S_n=X_1+X_2+...+X_n$ is the location of the particle at time n and has $\left[\widehat{f}(k)\right]^n$ as Fourier transform. Then $\frac{S_n}{\sqrt{n}}$ has Fourier transform $\left[\widehat{f}\left(\frac{k}{\sqrt{n}}\right)\right]^n$ and, using the central limit theorem, we have

$$\left[\widehat{f}\left(\tfrac{k}{\sqrt{n}}\right)\right]^n=\left(1-\tfrac{k^2}{n}+\cdots\right)^n\to e^{-k^2},\quad n\to\infty. \tag{8}$$

Inverting the Fourier transform, we obtain

$$\frac{S_n}{\sqrt{n}}\to\frac{1}{\sqrt{4\pi}}e^{-x^2/4}. \tag{9}$$

In like manner, expanding the time scale by a factor $c>0$,

$$\widehat{\frac{S_{[ct]}}{\sqrt{c}}}\to e^{-k^2t}=e^{t(ik)^2},\quad c\to\infty \tag{10}$$

($[ct]$ is the integer part of ct). But $e^{t(ik)^2}=\widehat{u}(k,t)$, from the diffusion equation. It results

$$\frac{S_{[ct]}}{\sqrt{c}}\to u(x,t)=\frac{1}{\sqrt{4\pi t}}e^{-x^2/4t}, \tag{11}$$

as $c\to\infty$. If $\mu_1=0$ and $\widehat{f}(k)=1+(ik)^{\alpha}+\cdots$ is the Fourier transform of a typical random variable which has a stable probability distribution with characteristic exponent α then

$$\widehat{\frac{S_n}{n^{1/\alpha}}}\to e^{-(ik)^{\alpha}}, \tag{12}$$

when $n \to \infty$.

Inverting this Fourier transform we obtain an α - stable density. Roughly speaking, this means that the probability of jumping a distance greater than r falls off like r^{α}. Moreover,

$$\frac{\widehat{S_{[ct]}}}{c^{1/\alpha}} \to e^{t(ik)^{\alpha}}, \tag{13}$$

as $c \to \infty$. The limiting process is called a Levy motion and its probability densities solve the fractional diffusion equation

$$\frac{\partial u}{\partial t}(x,t) = \frac{\partial^{\alpha} u}{\partial x^{\alpha}}(x,t) \tag{14}$$

for $0 < \alpha \leq 2$.

In classical diffusion, particles spread in a normal bell-shaped pattern according to Gaussian probability distribution. Anomalous diffusion occurs when the growth rate or the shape of the particle distribution is different from Gaussian. Very large particle jumps are associated with fractional derivatives in space. Very long waiting times lead to fractional derivatives in time. Empirical evidence shows that the waiting time between jumps is correlated with the ensuing size of the particle jumps. For these models, the limiting particle distribution is governed by a fractional differential equation, called coupled space-time fractional derivative equation.

COUPLED SOLAR WIND-MAGNETOSPHERE SYSTEM AND MATHEMATICAL MODELS

Ever since it was clear that Earth's magnetosphere is influenced by the Sun, important effort has been devoted to establishing the relationship between fluctuations in the energy delivered by the solar wind to the magnetosphere and variations in the magnetospheric response. The problem is to reflect connections and similarities between processes occurring on different spatial and temporal scales, which are revealed both in solar wind and in the magnetosphere. The recent studies of multiscale behavior in the coupled solar wind-magnetosphere system have based on two approaches: observational data from different sources which analyze the scaling properties and mathematical models (such as Fokker-Plank type equations) which interpret these processes.

Using [1], let $x(t)$ be a physical variable of interest, for example, the density of ions, or a component of the magnetic field, or the plasma current in a specified direction. For a system that exhibits self-similarity, the probability distribution function (*pdf*) of a variable $x(t)$ can be expressed as $F(x,t)dx = F(x/t^{w})dx/t^{w} = F(z)dz$, where $z = x/t^{w}$. The exponent w is to be determined from the dependence of the first moment on t as

$$\mu_1(t) = \int |x| F(x,t)\,dx = const \cdot t^{w} \tag{15}$$

and the normalization condition $\int F(x,t)\,dx = 1$. The second moment is $\mu_2(t) = const \cdot t^{2w}$.

For normal diffusion,

$$F(x,t) = t^{-1/2}\tfrac{1}{2\sqrt{\pi}}e^{-x^2/4t}, \quad -\infty < x < \infty, \quad t \geq 0, \tag{16}$$

we have the similarity variable $x/t^{1/2}$ and $2w = 1$. The moments of even order are:

$$\mu_{2n}(t) = \int_{-\infty}^{\infty} x^{2n} F(x,t)\,dx = \tfrac{(2n)!}{n!}t^n, \quad n = 0,1,2,..., \quad t \geq 0. \tag{17}$$

The variance $\sigma^2 = \mu_2(t) = 2t$ is proportional to the first power of time. Different values of $2w$ (for example $2w = 1$) can result from the Fokker-Plank type equation

$$\frac{\partial F}{\partial t}(x,t) = \frac{\partial}{\partial x}D(x)\frac{\partial F}{\partial x}(x,t), \tag{18}$$

where $D(x)$ describes nonuniformity of the diffusion coefficient ([1]). The main feature of a diffusion process is that $F(z)$ decays exponentially as z converges to infinity and all moments are finite $\mu_n(t) < \infty, \ 0 < t < \infty$. The same situation is valid for the so-called sub-diffusion described by

$$\tfrac{\partial^\beta F}{\partial t^\beta}(x,t) = \tfrac{\partial}{\partial x}D(x)\tfrac{\partial F}{\partial x}(x,t), \quad 0 < \beta < 1. \tag{19}$$

The evolution of $F(x,t)$ in the presence of multiscale processes is given by

$$\tfrac{\partial^\beta F}{\partial t^\beta}(x,t) = D_{\alpha\beta}\tfrac{\partial^\alpha F}{\partial x^\alpha}(x,t), \quad 0 < \alpha \leq 2, \quad 0 < \beta \leq 2 \tag{20}$$

([1], [2]). In the case of $\alpha = 2$ and $\beta = 1$,

$$\frac{\partial u}{\partial t}(x,t) = D\frac{\partial^2 u}{\partial x^2}(x,t), \tag{21}$$

$D_{\alpha\beta} = D$ is the diffusion coefficient. From now on, for the sake of convenience, we agree to put $D_{\alpha\beta} = 1$. Unlike the usual diffusion equations, the solution of this equation yields non-convergent moments, showing its multiscale features. For $\beta = 1$ and $0 < \alpha < 2$ this equation describes the Levy process.

FORMULAS AND NUMERICAL EXPERIMENTS

We denote with $F_{\alpha,\beta}(x,t)$ the Green function (or fundamental solution) of the Cauchy problem

$$\tfrac{\partial^\beta F}{\partial t^\beta}(x,t) = \tfrac{\partial^\alpha F}{\partial x^\alpha}(x,t), \quad 0 < \alpha \leq 2, \quad 0 < \beta \leq 2, \quad -\infty < x < \infty, \quad t \geq 0, \tag{22}$$

with $u(x,0) = \delta(x)$. Introducing the similarity variable x/t^γ, we can write $F_{\alpha,\beta}(x,t) = t^{-\gamma}R_{\alpha,\beta}(x/t^\gamma)$, $\gamma = \beta/\alpha$. Thus we express $F_{\alpha,\beta}(x,t)$ in terms of a function of a single variable, the reduced Green function. For example, the Green function of the standard diffusion $\alpha = 2$, $\beta = 1$,

$$F_{2,1}(x,t) = t^{-1/2}\frac{1}{2\sqrt{\pi}}e^{-x^2/4t} \tag{23}$$

has the similarity variable $x/t^{1/2}$. It results

$$R_{2,1}(x) = \frac{1}{2\sqrt{\pi}}e^{-x^2/4}. \tag{24}$$

We have some relevant particular cases of the equation

$$\frac{\partial^\beta F}{\partial t^\beta}(x,t) = \frac{\partial^\alpha F}{\partial x^\alpha}(x,t): \tag{25}$$

a) $0 < \alpha < 2$, $\beta = 1$ (space-fractional diffusion);
b) $\alpha = 2$, $0 < \beta < 2$ (time-fractional diffusion);
c) $0 < \alpha = \beta \leq 2$ (neutral-fractional diffusion), for which the fundamental solution can be interpreted as a probability density function (pdf).

For the standard diffusion equation it is well known that the fundamental solution of the Cauchy problem is the spatial probability density function (pdf) for the Gaussian distribution, whose variance is proportional to time. Indeed, we have

$$F_{2,1}(x,t) = t^{-1/2}\frac{1}{2\sqrt{\pi}}e^{-x^2/4t} = \frac{1}{2\sqrt{\pi}\sigma}e^{-x^2/(2\sigma^2)} = p_G(x,\sigma), \tag{26}$$

where $\sigma^2 = 2t$. $p_G(x,\sigma)$ denotes the Gauss pdf whose moment of the second order, the variance, is σ^2.

Let now $\psi_\alpha(k)$ be the characteristic function $\psi_\alpha(k) = (ik)^\alpha = |k|^\alpha e^{isign(k)\alpha\pi/2}$. We observe $\widehat{F_{\alpha,1}}(k,t) = e^{-t\psi_\alpha(k)}$. Then the solution of the space-fractional diffusion equation can be interpreted as a pdf. We have

$$F_{\alpha,1}(k,t) = t^{-1/\alpha}R_{\alpha,1}\left(x/t^{1/\alpha}\right). \tag{27}$$

The stable densities admit a representation in terms of elementary functions only in some particular cases. For example

$$R_{2,1}(x) = \tfrac{1}{2\sqrt{\pi}}e^{-x^2/4}, \quad R_{1,1}(x) = \tfrac{1}{\pi(x^2+1)}. \tag{28}$$

Generally, by inverting the Fourier transform, we obtain representations of the stable pdf's in terms of convergent and asymptotic power series ([3], [4]).

We consider $x > 0$ since the evaluations for $x < 0$ can be obtained using the symmetry relation: $R_{\alpha,1}(-x) = R_{\alpha,1}(x)$.

The convergent expansions are:

i) if $0 < \alpha < 1$, then

$$R_{\alpha,1}(x) = \tfrac{1}{\pi x}\textstyle\sum_{n=1}^{\infty} \tfrac{(-1)^{n+1}}{x^{\alpha n}}\tfrac{\Gamma(1+n\alpha)}{n!}\sin\left(\tfrac{n\pi\alpha}{2}\right), \quad x > 0; \tag{29}$$

ii) if $1 < \alpha < 2$, then

$$R_{\alpha,1}(x) = \frac{1}{\pi x}\sum_{n=1}^{\infty}(-1)^{n+1}x^{n}\frac{\Gamma(1+n/\alpha)}{n!}\sin\left(\frac{n\pi}{2}\right), \quad x > 0; \tag{30}$$

iii) if $\alpha = 1$, then $R_{1,1}(x) = \frac{1}{\pi(x^2+1)}$, $-\infty < x < \infty$;

iv) if $\alpha = 2$, then $R_{2,1}(x) = \frac{1}{2\sqrt{\pi}}e^{-x^2/4}$, $-\infty < x < \infty$;

v) if $\alpha = \beta$ then

$$R_{\alpha,\alpha}(x) = \frac{1}{\pi x}\sum_{n=0}^{\infty}(-1)^{n+1}x^{\alpha n}\sin\left(\frac{n\pi\alpha}{2}\right), \quad 0 < x < 1 \tag{31}$$

and

$$R_{\alpha,\alpha}(x) = \frac{1}{\pi x}\sum_{n=0}^{\infty}\frac{(-1)^{n+1}}{x^{\alpha n}}\sin\left(\frac{n\pi\alpha}{2}\right), \quad x > 1; \tag{32}$$

vi) for $\alpha < \beta$ we have

$$R_{\alpha,\beta}(x) = \frac{1}{\pi x}\sum_{n=1}^{\infty}\frac{(-1)^{n+1}}{x^{\alpha n}}\frac{\Gamma(1+n\alpha)}{\Gamma(1+n\beta)}\sin\left(\frac{n\pi\alpha}{2}\right), \quad x > 0; \tag{33}$$

we observe that for $\beta = 1$ and $0 < \alpha < 1$, we recover the series giving the representation of the stable density;

vii) if $\alpha = 2$ and $0 < \beta < 2$ then

$$R_{2,\beta}(x) = \frac{1}{\pi}\sum_{n=1}^{\infty}\frac{(-1)^{n-1}}{(n-1)!}x^{n-1}\Gamma\left(\frac{n\beta}{2}\right)\sin\left(\frac{n\pi\beta}{2}\right), \quad x > 0. \tag{34}$$

Thus we enable to plot the probability densities for different values of the α and β (we consider the independent variable x in the range [-5,5] and, in order to point out the tails, we prefer the logarithmic scale for the ordinate). Analyzing the graphs from Figures 1-2 we see the absence of heavy tails at $R_{2;1}(x)$, $R_{1.8;1}(x)$, $R_{2;1.5}(x)$, $R_{2;0.5}(x)$. $R_{2;0.5}(x)$ is an anomalous slow difussion ($0 < \beta < 1$) and $R_{2;1.5}(x)$ is an anomalous fast difussion ($1 < \beta < 2$).

The graphs of $R_{0.2;0.6}(x)$ and $R_{0.5;1}(x)$ present heavy tails (see Figure 3). Moreover, we observe that the moments are non-convergent. For example, if $\alpha = 0.5$ and $\beta = 1$ then the moments μ_p are finite if the integral is convergent, i.e. $\alpha + 1 - p > 1$, hence $\alpha > p$. The same conclusion is obtained from a set of data (example: solar wind electric field) which is used to compute the probability distribution function $p(n) = \frac{c}{n^{1+d}}$ where the exponent d depends on the type of the data and on the interval ([1]).

CONCLUSIONS

The fractional diffusion equations, which are used to represent the complexity, provide a suitable mathematical framework for the multiscale behavior. Unlike the usual diffusion equations, the solutions of these equations yield non-convergent moments, showing its multiscale features. Numerical solutions of these equations are used to analyze the nature of the equations by computing the moments, which are divergent. On the other hand the equation with space dependent diffusion coefficients yield convergent moments, indicating Gaussian type solutions and absence of heavy tails typically associated with multiscale behavior.

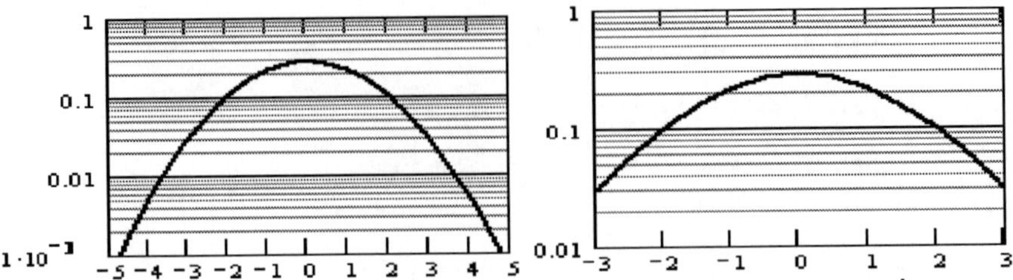

FIGURE 1. The graphs of $R_{2;1}(x)$ (left) and $R_{1.8;1}(x)$ (right)

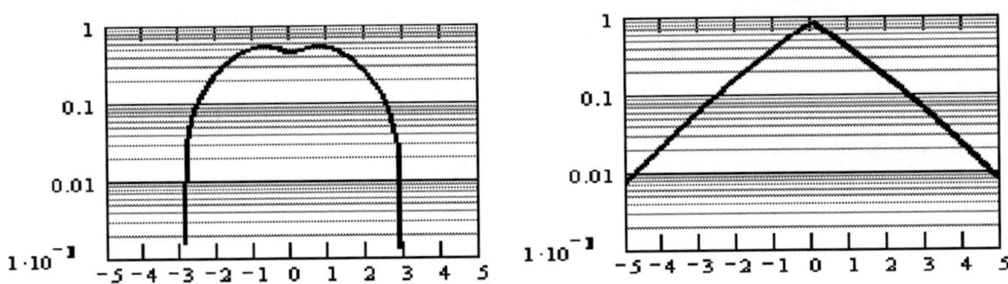

FIGURE 2. The graphs of $R_{2;1.5}(x)$ (left) and $R_{2;0.5}(x)$ (right)

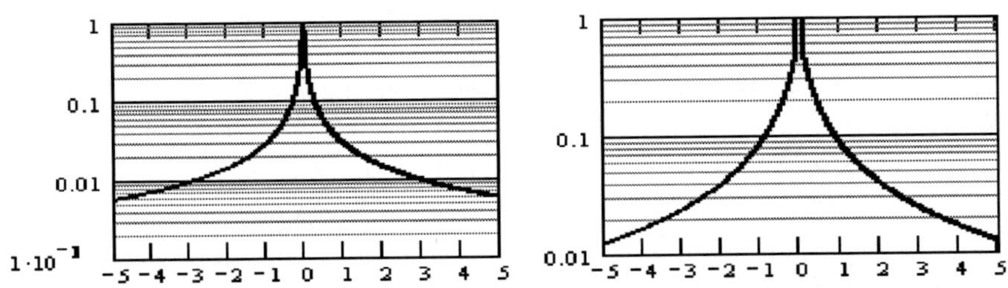

FIGURE 3. The graphs of $R_{0.2;0.6}(x)$ (left) and $R_{0.5;1}(x)$ (right)

REFERENCES

1. G. M. Zaslavsky, P. N. Guzdar, M. Edelman, M. I. Sitnov and A. S. Sharma, Multiscale behavior and fractional kinetics from the data of solar wind - magnetosphere coupling, arXiv:physics/0511096.
2. G. M. Zaslavsky, Fractional kinetic equation for Hamiltonian chaos, *Physica D: Nonlinear Phenomena*, **76**, 110-122 (1994).
3. W. Feller, An introduction to probability theory and its applications, New York, Wiley (1971).
4. R. Gorenflo and F. Mainardi, Random walk models for space-fractional diffusion processes, *Fractional Calculus and Applied Analysis*, **1**, 167-191 (1998).

Nonstandard Representations of Processes in Dynamical Systems

Žarko Mijajlović*, Nadežda Pejović† and Slobodan Ninković**

*Faculty of Mathematics, University of Belgrade, Serbia, e-mail: zarkom@matf.bg.ac.yu
†Faculty of Mathematics, University of Belgrade, Serbia, e-mail: nada@matf.bg.ac.yu
**Astronomical Observatory in Belgrade, Serbia, e-mail: sninkovic@aob.bg.ac.yu

Abstract. The aim of this paper is to present an alternative description, based on the non-standard analysis, of certain processes in dynamical systems. Some examples of applications will be given, including differential equations and geometry. In particular, bifurcations in perturbed systems are studied which may be applied to the study of a body having a complex magnetic field excited by strong microwave fields. A possible application might include modeling of certain facts of magnetic fields of Sun-like stars. The Karamata notion of a slowly varying function is used for analyzing such processes in infinitesimal neighborhoods (monads). These results, we believe, might give at least new insights into classical problems of dynamical systems.

Keywords: non-standard analysis, dynamical system, bifurcation, slowly varying function
PACS: 02.10.De, 98.80.Jk

INTRODUCTION

Infinite quantities reflect certain non-archimedean properties of underlying structure. If $^*\mathbf{R}$ is a non-standard model of real numbers, then infinitesimals (or infinite numbers) in $^*\mathbf{R}$ arise from the non-archimedean property of the order of $^*\mathbf{R}$. Thus, $^*\mathbf{R}$ is the appropriate structure for modelling physical phenomena referring to infinite and infinitesimal notions. The non-standard analysis has been presented in some extent in explaining certain phenomena in quantum physics and statistical physiscs (S. Albeverio, J.E. Fenstad, T. Lindstrom, see [**1**]).

In writing this paper we had in mind several goals:

1. To promote the use of the non-standard (Leibnitz's) analysis in the study of certain physical phenomena, in particular in astronomy.
2. Use of the notion of the Karamata slowly varying functions in thy study of bifurcations in infinitesimal neighborhoods - monads.
3. Formulation of the notions of bifurcations in terms of the non-standard analysis.

NON-STANDARD ANALYSIS

The classical approach to physics is based on the mathematics over **R**, the field of real numbers, or **C**, the field of complex numbers. Both these structures are archimedean, i.e. they do not admit explicitly infinite quantities.

CP934, *Flows, Boundaries, Interactions*
edited by C. Dumitrache, V. Mioc, and N. A. Popescu

In discussing the real line **R** we have no way of knowledge what a line in physical space is really like. It might be like the real line **R**, the hyperreal line $^*\mathbf{R}$ or neither. However, in applications of the mathematical analysis it is helpful to imagine a line in physical space as $^*\mathbf{R}$. The hyperreal line is, like the real line, a useful mathematical model for a line in physical space. The non-standard analysis is based on properties of $^*\mathbf{R}$ and the transfer principle (Łoś theorem) which transfer propositions on elements between $^*\mathbf{R}$ and **R**.

The origin of infinitesimals goes back to Leibnitz who assumed "infinitely small numbers", and he performed usual algebraic operations over them in the same way as he did with real numbers. In particular, each positive infinitesimal ε is lesser than any standard (real) positive number, while $1/\varepsilon$ is greater than any standard positive number. His infinitesimal calculus was one of the most useful achievements in mathematics but it was not founded. Weierstrass founded analysis by introducing the $\varepsilon - \delta$ formalism, but since then the notion of infinitesimal was expelled from mathematics. In fact, the zero-sequences took the role of infinitesimals, however the problem was that even simple arithmetical operations, such as division, could not be performed on them. Abraham Robinson gave the true foundation of the Leibnitz analysis in his seminal paper [2] and this discipline, or better to say this approach to mathematics, since then is called non-standard analysis. We won't go further here into historical details, neither in foundation, but we shall give a short review of some of basic notions of the non-standard analysis.

Let **R** be the ordered field of real numbers expanded by the collection S of functions and relations that we intend to study. By $^*\mathbf{R}$ we shall denote an $\aleph_1$-saturated non-archimedean elementary extension of **R**. Then the extension function $*$ assigns to every object $X \in S$ a non-standard counterpart $^*X$. All images $^*X$ are called internal objects. For example, if N is the set of natural numbers, then $^*N$ is the set of non-standard natural numbers and $^*N$ is an internal set. The map $*$ adds new elements to every infinite X, i.e. $^*X \backslash X$ is nonempty and elements of $^*X \backslash X$ are called non-standard. For example, elements of $^*N \backslash N$ are called non-standard natural numbers, while elements of $^*R \backslash R$ are called non-standard real numbers. The most important is the following transfer principle:

Leibnitz principle: *Every mathematical proposition that is true for finite (real) numbers is also true for the extended system (i.e. system with infinite numbers), and vice versa.*

A consequence of the Compactness Theorem is the fact that the described theory has a model. In particular, each model of that theory contains infinitesimals. We see that this model is in fact a non-archimedean real field but enriched with non-standard counterparts of notions of the mathematical analysis: elementary functions $\sin(x), \ln(x), \ldots$, sets: natural numbers **N**, integers **Z**, rational numbers **Q**, etc. We also see that the following extension principle holds:

Extension property: *Every function* $f : \mathbf{R} \longrightarrow \mathbf{R}$ *can be extended to the function* $^*f : {}^*\mathbf{R} \longrightarrow {}^*\mathbf{R}$ *which preserves all first order properties of* f.

For example, if $f(x) = \sin(x)$, $g(x) = \cos(x)$, since

$$sin(x+y) = \sin(x)\cos(y) + \cos(x)\sin(y),$$

the same identity holds for $^*f(x) = {}^*\sin(x)$ and $^*g(x) = {}^*\cos(x)$. Something similar is true for analytical continuations of real functions, but only for identities. In the non-standard analysis all first-order properties are preserved, including monotonicity, the properties of zeros, etc. It is customary that the asterisk * is omitted in the case of elementary functions. So, $\sin(x)$ will denote ${}^*\sin(x)$ if $x \in {}^*\mathbf{R}$, otherwise it is the standard sine function.

The best of all is that we can do the same with more complex structures. For example, to construct the non-standard enlargement of any infinite structure: complex numbers $\mathbf{C}$, the space of real sequences $\mathbf{R}^{\mathbf{N}}$, the space of real functions $\mathbf{R}^{\mathbf{R}}$, each having the metric according to our choice; then infinite functional, geometrical and topological spaces. This construction simply allows us non-standard but consistent mathematics. The Leibnitz transfer principle enables one to translate theorems from non-standard universe to the standard one. In particular, the Cauchy transfer principle is useful in such translations:

The Cauchy Principle *Let $\varphi(x)$ be an internal formula. Then: If $\varphi(x)$ holds for each infinitesimal x, then there is $r \in \mathbf{R}, r > 0$ so that $\varphi(x)$ holds in $\mathbf{R}$ for all $x \in \mathbf{R}, |x| < r$.*

An element $a \in {}^*\mathbf{R}$ is *finite* if there is a positive integer n such that $-n < a < n$. By ${}^*\mathbf{R}_{\text{fin}}$ we shall denote the set of all finite elements of ${}^*\mathbf{R}$. The *galaxy* of a is the set $g(a)$ of all non-standard reals b such that $a - b$ is finite. In particular, ${}^*\mathbf{R}_{\text{fin}} = g(0)$. The mapping $\mathbf{st} : {}^*\mathbf{R}_{\text{fin}} \longrightarrow \mathbf{R}$ (*standard part*) defined by $\mathbf{st}(x) = \sup\{y \in R \mid y \leq x\}$ is an epimorphism. In particular, ${}^*\mathbf{R}_{\text{fin}}/\ker(\mathbf{st}) \cong \mathbf{R}$.

An *infinitesimal* is each finite ε such that $\text{st}(\varepsilon) = \text{st}(0) = 0$. The *monad* of 0 is the set $\mu(0)$ of all infinitesimals. Notice that $\mu(0)$ is closed under addition and multiplication. Further, we say that a and b are infinitely close (denoted by $a \approx b$) if $a - b \in \mu(0)$. In fact, $\mu(0)$ is a kernel of epimorphism **st** and it is a maximal ideal of ${}^*\mathbf{R}_{\text{fin}}$. The other monads are obtained by translations, i.e. $\mu(a) = a + \mu(0)$.

By use of the homomorphism **st** one can replace the ε-δ formalism by algebraic identities.

For detailed development of non-standard analysis, one may consult [7]. Notions and tools of model theory (the part of mathematical logic) used in the development of the non-standard analysis one can find, for example, in [6].

Examples

Here we give several examples of basic mathematical notions expressed in non-standard terms and non-standard solutions of some classical problems.

1. Let S be an infinite subset of $\mathbf{R}$. Then ${}^*S \setminus S$ is also infinite.
2. If ε is an infinitesimal, then $\text{st}(a + \varepsilon) = \text{st}(a)$.
3. $f : \mathbf{R} \longrightarrow \mathbf{R}$ is continuous iff for all $a \in {}^*\mathbf{R}_{\text{fin}}$ $\text{st}({}^*f(a)) = f(\text{st}(a))$, i.e. for every infinitesimal ε there is an infinitesimal η such that ${}^*f(a + \varepsilon) = f(a) + \eta$.
4. Let $f : \mathbf{R} \longrightarrow \mathbf{R}$ be a differentiable function and let $\varepsilon \neq 0$ be an infinitesimal. Then $f'(x) = \text{st}\left(\frac{{}^*f(x+\varepsilon)-f(x)}{\varepsilon}\right)$. For example, $(x^2)' = \text{st}\left(\frac{(x+\varepsilon)^2-x^2}{\varepsilon}\right) = \text{st}(2x + \varepsilon) = 2x$.
5. $\int_0^1 f(x)dx = \text{st}(\frac{1}{H}\sum_{i=0}^{H} {}^*f(\frac{i}{H}))$, where $H \in {}^*N \setminus N$ and f is continuous.

6. **Karamata slowly varying and regularly varying functions.** In certain dynamical systems there are slowly varying processes, or it is needed a detailed analysis of the process in the neighborhood of the particular points. Examples of this kind include critical (equilibrium) points of a dynamical system generated by an autonomous system of ordinary differential equations (ODEs). It appears that Karamata slowly varying and regularly varying functions might be useful in these analysis. They are defined as follows:
Slowly varying functions: $\lim_{x\to+\infty} L(tx)/L(x) = 1, \quad t > 0,$
$\text{Examples}: \ \ln x, \ \ln(x)^2, \ \ln^{\alpha}(x) \ \ldots$
Regularly varying functions: $\lim_{x\to+\infty} R(tx)/R(x) = t^{\rho}, \quad t > 0,$
$\text{Examples}: \ x, x^2, \ x^{\alpha}, \ x^{\alpha}\ln^{\beta} x, \ \ldots$
L and R being continuous and positive on a positive half-axis. The constant ρ is called the index of regularity. This two definitions imply that any regularly varying function has the form $R(x) = x^{\rho}L(x)$. Starting from these two simple definitions, Karamata developed the whole new theory of slowly and regularly varying functions which has included the majority of the most important properties of these functions. The definitions given above appeared in 1930 in the seminal paper [1]. The intent of his paper was to generalize the Tauberian conditions in some inverse theorems of the Tauberian type for the Laplace transform. But it was soon realized that those functions could be successfully applied to many branches of mathematical analysis, probability theory and even in astronomy, [9].
Jovan Karamata (1902-1967) is one of the most known Serbian mathematician and his discovery of these notions is his most famous result.
The non-standard description of slowly varying functions is as follows: For all positive infinite $H \in {}^*\mathbf{R}$, $L(tH)/L(H) \approx 1, t > 0$.
7. **Dirac delta function**. Let $a(t) = e^{-1/(1-|t|^2)}$ if $|t| < 1$, $a(t) = 0$ otherwise. This is a simple variation of Cauchy's flat function, and it belongs to the space $\mathscr{E}^{\infty}$ of infinitely many differentiable functions. Let ε be a positive infinitesimal, and let $b(t) = a(t/\varepsilon)$. Finally, let $k = \int_{-\infty}^{\infty} b(t)dt$ and let $\delta(t) = b(t)/k = a(t/\varepsilon)/k$. Then $\delta(t)$ belongs to ${}^*\mathscr{E}^{\infty}$, it is positive, and has integral one. In fact, it is what is expected, $\delta(t)$ is a finite compact distribution and it has all properties attributed to the Dirac function.
8. The next example is from statistical physics. Anderson [4] gave a non-standard description of random walk and Brownian motion: Let $H \in^* N$ be infinite, $\Delta t = H^{-1}$ and $T = \{k\Delta t \ |k \in^* N, k \leq H\}$. Then T represents time and $B(\omega,t) = \sum_{s=0}^{t-\Delta t} \omega(s)\sqrt{\Delta t}$, $\omega{:}T \to \{1,-1\}$, is a hyperfinite random walk, moving by $\sqrt{\Delta t}$ right or left at each infinitesimal interval Δt. Then, $b(\omega, \mathrm{st}(t)) = B(\omega,t)$ is a realization of the Brownian motion on $\mathbf{R}$.
9. The next examples are from geometry, and they could explain certain phenomena in astronomy.
Tiling the Euclidean plane (patterns, e.g. dominoes): if there is a covering of each square (circle), prove that we can cover the entire plane. One solution goes like this: by the extension principle, we can find the covering of a square with edges (radius) with the length H, $H \in {}^*N \setminus N$. This particular cover induces the covering

of the entire Euclidean plane. Here one can recognize the foundation of one of the basic cosmological principles: homogeneity of the Universe. From the preceding example it follows that from the mathematical point of view at least it is consistent to assume so. This example is connected with famous Hao Wang problem of tiling the plain with dominos (Wang tiling), [3].
Is every parabola ellipse? Let E be an ellipse touching the y-axis and having focuses in $(0,1)$ and $(0,H)$ where $H>0$ is an infinite real number. Then the set P of all standard points of E, i.e. points laying in **R**, are points of loci of an "ordinary" (standard) parabola, $y^2=4x$. In fact, P is the envelope of the family of all ellipses touching y-axis and having one focus in $(0,1)$, the other one in $(0,b)$, $b>1$ is a (standard) real number. Here one can recognize the difficulty in determination of the nature of comet orbits with a remote second focus. This example shows that every comet's orbit which we measured (observed) as parabolic, actually is elliptical. But, it's second focus is excessively far away to measure it.
The cell structure of the Universe. The space $^*R^n$ is partitioned into monads, $^*\mathbf{R}^n_{fin}=\bigcup_{r\in R^n}\mu(r)$. In [10] authors discussed a mathematical model of quantized time and space of the Universe partitioned into cells. In addition, there an interpretation of the Planck distance and Planck time is given, and some conclusions on physics what it would be like below these bounds, as well.

Non-standard methods in dynamical system

The non-standard analysis introduces new concepts, which do not exist in the classical analysis. For example, $^*\mathbf{R_{fin}}$, the finite points of $^*\mathbf{R}$, is partitioned into monads: $^*\mathbf{R_{fin}}=\bigcup_{\mathbf{r}\in\mathbf{R}}\mu(\mathbf{r})$ We can think about $\mu(r)$ as the "cloud" of infinitely closed points to the standard real number r. Thus, if **R** is a counterpart of time-line, then it is sound to suppose that the time is quantized into infinitely small parts (monads), each part having rather rich structure. Similar observation could be asserted for the space $\mathbf{R^n}$ (see the last example in the previous section). Accordingly, notions of dynamical systems as described expressed in $\varepsilon-\delta$ formalism have natural and more intuitive description.
We consider the ordinary differential equation:

$$(1)\qquad \dot{x}=dx/dt=f(x,\mu),\quad \mu\in\mathbf{R},\ f\in C^2(\mathbf{R}),$$

where μ is a parameter. The variable t represents time, i.e. $t\in R$, and μ is a constant parameter if it does not depend on time t.
A value μ_0 for which the flow of (1) is not structurally stable is called a *bifurcation.* Simply saying, (1) undergoes a bifurcation for μ_0 if the critical point x_c of (1) behavior changes. To remind the reader, the critical points of (1) are the roots of $f(x,\mu_0)=0$, i.e. $f(x_c,\mu_0)\equiv 0$, so x_c is also the fixed point of (1).
The critical point x_c is *stable* if it captures all paths (trajectories, orbits) in the phase space $(x,\dot{x})$ of (1) crossing close enough to x_c. More formally, if the path $p(t)$ crosses close enough to x_c, i.e. at the distance less than given $\varepsilon>0$, then for

every neighborhood V_{x_c} of x_c there is t_0 such that for any $t \geq t_0$, Vx_c contains all points of $p(t)$.
If x_c is not stable, than we say x_c is *unstable* and it is the source of bifurcation.
Suppose there is an external influence represented by μ acting on the system modeled by (1) varying in time t. Then $\mu = \mu(t)$, and (1) is not anymore an autonomous system.
We will discuss *delayed dynamic bifurcation* which may occur if $\mu(t)$ is slowly varying function crossing the bifurcation point x_c. The slowly varying function we shall define in non-standard terms.
Dynamical system is *delayed* if there are initial values (t_0, x_0), $x_0 \neq x_c$, and time interval $(0, t_j)$ such that every solution x_j of (1) with $x_j(t_0) = x_0$ is crossing arbitrarily close to x_c.
We consider also delayed dynamic systems with perturbation. Perturbation could be, for example, constant or noise. It appears that if perturbation is small, then delay survives, Walter.
Non-standard formulation of stability. The critical point x_c of (1) in 1.3 is stable if any path $p(t)$ of (1) crossing close to x_c is captured at infinity by the monad $\mu(x_c)$ of x_c. That is, for all infinite time T, $p(T)$ is inside $\mu(x_c)$. Therefore, x_c is unstable if for some infinite time T, the path $p(T)$ is outside of $\mu(x_c)$. This is exactly what we mean by bifurcation. Small changes in the process driving the dynamical system given by (1), produces in certain time t_0 the instability of the system.

From the preceding example we see that it is worth to study path in the infinity and the monad $\mu(x_c)$. For this purpose we introduce the notion of μ-slowly varying function. Let $f{:}R^2 \longrightarrow R$. Then the function f is μ-slowly varying at point $a \in R$ if for every $x \approx a$, ${}^*f(\varepsilon, t) \in \mu(0)$, $t \in {}^*\mathbf{R}_{fin}$. An example of μ-slowly varying function at x_c is

$$\tau(\varepsilon, t) = \mu_0 + \varepsilon(t - t_0). \tag{2}$$

Other examples of μ-slowly varying functions are $\tau(\varepsilon, t) = \mu_0 + \varepsilon \sin(t)$ and $\tau(\varepsilon, t) = \mu_0 + \sin(\varepsilon t)$. These functions might be useful in dynamical systems with perturbation.

Local time and slowly varying functions

Good source of μ-slowly varying functions are the Karamata slowly varying functions and regularly varying functions. μ-slowly varying functions are easily obtained from Karamata types functions by simple transformations. If $K(t)$ is Karamata slowly varying function and $k(x,t) = (K(x,t) - K(x))/K(x)$, then for all positive infinite H, ${}^*k(H,t) \in \mu(0)$ for all finite t. Thus, $\tau(\varepsilon, t) = k(1/\varepsilon, t)$ is the μ-slowly varying function. Now we take the local time at x_c, $\tau(t) = \mu(t)$, for example taking $\mu(t)$ as in (2). Then (1) becomes:

$$\varepsilon dx/dt = f(x, \tau).$$

It should be noticed that if τ takes positive standard time, t must be infinite, exactly what we want, when the path $p(t)$ might enter $\mu(x_c)$ (and this is the case when x_c is stable).

Non-standard formulation of delayed dynamical system. Consider a system of type (2) with positive infinitesimal ε. For simplicity we shall assume that x_c is stable for $\tau < 0$ and unstable for $\tau > 0$. If there exist initial values (x_0, τ_0), $\tau_0 < 0, \tau_0 \not\approx 0$, and $x_0 \not\approx x_c$, such that the solution of (2) with $x(\tau_0) = x_0$ for some $\tau > 0, \tau \not\approx 0$ is satisfying:

$$x(\tau') \approx x_c\,, \quad \tau' \in (0,\tau)\,. \tag{3}$$

then it is a delayed bifurcation.
The solution of (2) may consist of an infinite family F of functions. Then it is easy to check that F is the set of bounded equicontinuous functions, therefore, according to the Ascoli theorem, there is a continuous function $g(t)$ and a sequence $f_n(t) \in F$ which converges point-wise to $g(t)$. In other words, the family F is glued (or attracted) to $g(t)$. Examples in [8] which arise in the study of high-power magnetic resonance are good illustration of the statement we just noted.

Conclusion

The non-standard analysis could be useful in studying certain problems in astrophysics and physics in general. Definition of many notions are more intuitive, while proofs and inferences are easier. In particular, applications of non-standard analysis in dynamical systems could be fruitful.

REFERENCES

1. J. Karamata, *Sur une mode de croissance reguliere des fonctions*, Mathematica, Cluj, 1930.
2. A. Robinson, *Non-Standard Analysis*, Proc. Roy. Acad. Amsterdam, ser. A, 64, 432-440, 1961.
3. H. Wang, *Proving theorems by pattern recognition, II*, Bell Systems Techical Jour., vol. 40, 1-41, 1961.
4. R.M. Anderson, *A non-standard representation for Brownian motion and Ito integration*, Isr. J. Math, 25, 15-46, 1976.
5. S. Albeverio et al, *Non-standard Methods in Stohastic Analysis and Mathematical Physics*, Academic Press, New York 1986.
6. C. C. Chang, H. J. Keisler, *Model Theory*, North–Holland, Amsterdam, 1990.
7. K. D. Stroyan, W. A. J. Luxemburg, *Introduction to the theory of infinitesimals*, Academic Press, New York, 1976.
8. T. Walter, Delayed bifurcations in perturbed system analysis of slow passage of Suhl-treshhold, in *Developments in nonstandard mathematics*, ed. N.J. Cutland at al, Longman, Harlow, 61-72, 1995.
9. I. Stern, *On Fractal Modeling in Astrophysics: The Effect of Lacunarity on the Convergence of Algorithms for Scaling Exponents*, Astronomical Data Analysis Software and Systems VI, ASP Conference Series, Vol. 125, 1997.
10. Z. Mijajlovic, N. Pejović, *Non-Archimedean Methods in Cosmology*, AIP Conference proceedings, vol. 895, New York, 317-320, 2007.

STARS AND GALAXIES

Cosmological Flows

Marian Doru Suran

Astronomical Institute of the Romanian Academy, Bucharest, Romania, e-mail: suran@aira.astro.ro

Abstract. The aim of this paper is to present different types of flows in the Universe, from cosmological (global Universe evolution) to cosmical ones (local structures at the large scale in the Universe).

Keywords: Cosmology
PACS: 98.80.-k, 98.80Jk, 98.80Bp, 95.30

INTRODUCTION

Before discussing 'cosmological flows' we must define a couple of terms.

A basic cosmological scenario is described by a General Relativistic (GR) model which consists of:

- a geometry (a spacetime *M* of metric g_{ab}, from a Big Bang of FLWR type)
- a collection of matter-fields (the energy-momentum tensor T_{ab})
- a theory of the interaction between geometry and matter, in the form of the Einstein field equation (EFE): $G_{ab} \equiv R_{ab} - \frac{1}{2}Rg_{ab} + \Lambda g_{ab} = T_{ab}$;

with two additional conservation equations: $\nabla_a G^{ab} = 0, \nabla_a T^{ab} = 0$.

There are two meanings of the term 'flow' - mathematical 'flow' and physical 'flow'.

Flow (mathematics), the one-dimensional (continuous, differentiable) action of a group, usually arising in the solution of differential equations. In mathematics, a flow formalizes, in mathematical terms, the general idea of "a variable that depends on time" that occurs very frequently in engineering, physics and the study of ordinary differential equations. Informally, if *x(t)* is a coordinate of some system that behaves continuously as a function of *t*, then *x(t)* is a flow. More formally, a flow is the group action of a one-parameter group on a set.

In our case, the mathematical 'flow' must be associated with the evolution of space-time as a special case of a mathematical Lorentzian manifold. Evolutive examples include geodesic flow, Ricci flow (curvature flow), Hamiltonian flow , and Anosov flow (ergodic flow).

Flow (physics), another name for 'flux', which is the rate at which something travels through a given cross section. In fluid mechanics, the word 'flow' is often used to mean a complete description of the motion of a fluid.

A fluid is defined as a substance that continually deforms (flows) under an applied shear stress regardless of the magnitude of the applied stress. Fluids share the properties of not resisting deformation and the ability to flow.

CP934, *Flows, Boundaries, Interactions*
edited by C. Dumitrache, V. Mioc, and N. A. Popescu

The way in which the cosmological scenario is defined expresses the deep relations between *mathematics* $\leftarrow$ *cosmology* $\rightarrow$ *physics*:

- The evolving Universe as a topological space (manifold) in dynamical evolution - mathematical flows,
- Evolution of structures in the evolving Universe described as (M)HD flows of different constitutive fluids - physical flows.

In this mode, the philosophical problem of the unreasonable effectiveness of mathematics in natural sciences expresses the mode in which we perceive the Universe - a mathematical object filled by physical fluids. The relation between fluid $\Leftrightarrow$ geodesic flows expresses the idea that fluid motion (perfect fluid in adiabatic processes) is equivalent to the geodesic motion of fluid particles in (pseudo) Riemannian spacetimes.

The cosmological treatment involves:

- Study of the universe as an isolated system in the flow of time.
- Study of the evolution of the universe and structure formation (Big Bang model):
$$\begin{array}{ccc} QG & \rightarrow & GR \\ quantum fluctuations & \rightarrow & halos/galaxies \end{array}$$
- Study of the main evolutive mechanisms in universe:
  - *Expansion* (global, accelerated (dominance of vacuum/fields) or decelerated (dominance of radiation/matter)),
  - *Contraction* (local, matter fluids, gravity, linear $\rightarrow$ *nonlinear*)
- as expression of:
  - Global cosmological evolution of the fluids (linear motions, flows),
  - Local perturbed motion of cosmological fluids (linear and nonlinear motions, non homogeneities grows, flows, fluxes, structures formation).

COSMOLOGICAL FLOWS

In our treatment of GR we used the 1+3 formalism ([1]) which is based on:

- the congruence of the world lines defined at CMB,
- the existence of fundamental observers.

Related to the geometry of the spacetime we define the kinematics of the fundamental congruence which fix the cosmological evolution (the cosmological flow):

- velocity: $u^a \equiv \frac{dx^a}{d\tau}$;
- acceleration: $\dot{u}^a \equiv u^b \nabla_b u^a$;
- expansion: $\theta \equiv \nabla_a u^a; (\frac{1}{3}\theta \equiv \frac{\dot{a}}{a})$;
- shear:$\sigma_{ab} \equiv \tilde{\nabla}_{(a} u_{b)}; (\tilde{\nabla}_c F_{ab} \equiv h^d{}_c h^e{}_a h^f{}_b \nabla_d F_{ef})$;
- rotation:$\omega_{ab} \equiv \tilde{\nabla}_{[a} u_{b]}$
- projection: $h_{ab} = g_{ab} + u_a u_b$.

The energy-momentum tensor (the contribution of the different fluids and fields in the cosmological flow) is chosen by a general formula:

$$T_{ab} = \underbrace{\mu u_a u_b + \frac{1}{3}\mu h_{ab} + 2q^{(a}u^{b)} + \pi^{ab}} + \underbrace{\rho u^a u^b} + \underbrace{\phi_{,a}\phi_{,b} - g_{ab}[\frac{1}{2}\phi_{,c}\phi^{,c} + V(\phi)]}$$

where the first set of terms represents the contribution of the radiation, the second the contribution of the CDM dust and the third the contribution of a scalar field.

Different contributing terms to the flow are defined by the

- relativistic energy density: $\mu \equiv u^a u^b T_{ab}$;
- isotropic pressure: $p \equiv \frac{1}{3}h^{ab}u^a u^b T_{ab}$;
- anisotropic pressure: $\pi_{ab} \equiv T_{(ab)}$;
- energy flux (heat flow): $q_a \equiv T_{(ab)}u^b$;
- speed of the sound: $c_s^2 = (\frac{\partial p}{\partial \mu})_{|S=const}$.

The equations of state (EOS) for different fluids and different cosmological flow regimes and could be defined by:

- γ law type equation of state: $p = (\gamma - 1)\mu = w\mu$;
- perfect fluid: $\pi_{ab} = q^a = 0 \rightarrow p = 0, \dot{S} = 0$;
- dust: $p = 0$;
- barotropic fluid: $p = p(\mu)$;
- scalar fields: $\mu = -\frac{1}{2}\nabla_a\phi\nabla^a\phi + V(\phi), p = -\frac{1}{2}\nabla_a\phi\nabla^a\phi - V(\phi)$;
- Λ case: $\phi = \Lambda = const., V(\phi) = \phi$.

To these equations we add the relativistic thermodynamical equations:

- number density conservation: $\nabla_a(nu^a) = 0 \rightarrow \dot{n} + \theta n = 0 \Leftrightarrow nl^3 = \text{const.}$;
- first law of thermodynamics: $T\ \nabla_a S dx^a = \nabla_a[\frac{\mu}{n}]dx^a + p\nabla_a[\frac{1}{n}]dx^a \Rightarrow \dot{S} = \frac{1}{Tn}[\mu + \theta(\mu + p)]$;
- second law of thermodynamics: $S^a = Snu^a + \frac{R^a}{T}$, where the term R^a collects all the dissipative processes.

To obtain a useful form for the EFE equation: $G_{ab} = T_{ab} + \Lambda g_{ab}$, in the 1+3 formalism, we split the Weyl tensor: $R_{abcd} = C_{abcd} + g_{a[c}R_{d]b} - \frac{1}{3}Rg_{a[c}g_{d]b}$, in:

- the electrical part: $E_{ab} \equiv C_{abcd}u^c u^d$;
- the magnetic part: $H_{ab} \equiv \frac{1}{2}\eta_{acde}C^{cd}{}_{bf}u^e u^f$.

Corresponding, we can split EFE in the form:

EFE evolution equations:

- energy conservation equation:
  $\dot{\mu} + \theta(\mu + p) + \tilde{\nabla}_a q^a = -2\dot{u}_a q^a - \sigma_{ab}\pi^{ab}$,
- Raychaudhuri equation: $\dot{\theta} + \frac{1}{3}\theta^2 = \tilde{\nabla}_a\dot{u}^a + \dot{u}_a\dot{u}^a - 2\sigma^2 + 2\omega - \frac{1}{2}(\mu + 3p) + \Lambda$,
- momentum equation: $\dot{q}^{(a)} + \frac{4}{3}\theta q^a + \tilde{\nabla}^a p + \tilde{\nabla}_b\pi^{ab} = -\sigma^{ab} - (\mu + p)\dot{u}^a - \dot{u}_b\pi^{ab} - \eta^{abc}\omega_b q_c$,

- shear equation: $\dot{\sigma}^{\langle ab\rangle}+\frac{2}{3}\theta\sigma^{ab}=\tilde{\nabla}^{\langle a}\dot{u}^{b\rangle}+\dot{u}^{\langle a}\dot{u}^{b\rangle}-\sigma^{\langle a}{}_{c}\sigma^{b\rangle c}-\omega^{\langle a}\omega^{b\rangle}-E^{ab}+\frac{1}{2}\pi^{ab}$,
- rotation equation: $\dot{\omega}^{\langle a\rangle}+\frac{2}{3}\theta\omega^{a}=\frac{1}{2}\eta^{abc}\tilde{\nabla}_{b}\dot{u}_{c}+\sigma^{ab}\omega_{b}$,
- electric equation: $\dot{E}^{\langle ab\rangle}+\theta E^{ab}-\eta^{cd(a}\tilde{\nabla}_{c}H^{b)}{}_{d}+\frac{1}{2}\dot{\pi}^{\langle ab\rangle}=-\frac{1}{2}[\tilde{\nabla}^{\langle a}q^{b\rangle}+(\mu+p)\sigma^{ab}]+3\sigma^{\langle a}{}_{c}[E^{b\rangle c}-\frac{1}{6}\pi^{b\rangle c}]-\dot{u}^{\langle a}q^{b\rangle}-\frac{1}{6}\theta\pi^{ab}+\eta^{cd(a}[2\dot{u}_{c}H^{b)}{}_{d}+\omega_{c}(E^{b)}{}_{d}+\frac{1}{2}\pi^{b)}{}_{d})]$,
- magnetic equation: $\dot{H}^{\langle ab\rangle}+\theta H^{ab}+\eta^{cd(a}\tilde{\nabla}_{c}E^{b)}{}_{d}-\frac{1}{2}\eta^{cd(a}\tilde{\nabla}_{c}\pi^{b)}{}_{d}=3\sigma^{\langle a}{}_{c}H^{b\rangle c}+\frac{3}{2}\omega^{\langle a}q^{b\rangle}-\eta^{cd(a}[2\dot{u}_{c}E^{b)}{}_{d}-\omega_{c}H^{b)}{}_{d}-\frac{1}{2}\sigma^{b)}{}_{c}q^{d}]$.

EFE constraints:

- shear divergency: $\tilde{\nabla}_{b}\sigma^{ab}=\frac{2}{3}\tilde{\nabla}^{a}\theta-\eta^{abc}[\tilde{\nabla}_{b}\omega_{c}+2\dot{u}_{b}\omega_{c}]-q^{a}$,
- vorticity divergence: $\tilde{\nabla}_{a}\omega^{a}=\dot{u}_{a}\omega^{a}$,
- curl equation: $\eta^{cd(a}\tilde{\nabla}_{c}\sigma^{b)}{}_{d}-H^{ab}=2\dot{u}^{\langle a}\omega^{b\rangle}+\tilde{\nabla}^{\langle a}\omega^{b\rangle}$,
- electric divergence: $\tilde{\nabla}_{b}E^{ab}-3\omega_{b}H^{ab}=\frac{1}{3}[\tilde{\nabla}^{a}\mu-\theta q^{a}]+\frac{1}{2}[\sigma^{ab}q_{b}-\tilde{\nabla}_{b}\pi^{ab}]+\eta^{abc}[\sigma_{bd}H^{d}{}_{c}-\frac{3}{2}\omega_{b}q_{c}]$,
- magnetic divergence: $\tilde{\nabla}_{b}H^{ab}+3\omega_{b}E^{ab}=\frac{1}{2}\omega_{b}\pi^{ab}-(\mu+p)\omega^{a}-\eta^{abc}[\frac{1}{2}\tilde{\nabla}_{b}q_{c}+\sigma_{bd}(E^{d}{}_{c}+\frac{1}{2}\pi^{d}{}_{c})]$.

NONPERTURBATIVE FLRW COSMOLOGICAL FLOW

We will apply the cosmological flow in the case of the Big Bang universe model.

The Big Bang theory depends on two major assumptions:

- the universality of physical laws: the mathematical effectiveness of the physical laws in description of the universe;
- the cosmological principle (combined with the Copernician principle): the universe is homogeneous and isotropic.

The Friedmann-Lemaître-Robertson-Walker (FLRW) metric (supplied with a Λ term) is an exact solution of the Einstein field equations of general relativity; it describes in a natural mode a homogeneous, isotropic universe in evolution (expanding or contracting flows). From a mathematical point of view, the homogeneity of the spacetime is defined if the associated isometry group acts transitively and globally upon that spacetime; and the isotropy is defined if the isometry groups for each point in the spacetime act transitively.

The resulting cosmological model is an Friedmann-Lemaitre-Robertson-Walker (FLRW) cosmological model in a ΛCDM scenarios by:

- metric (the FLRW metric, the generic one under the assumption of homogeneity and isotropy and that the spatial component of the metric can be time dependent): $ds^{2}=-dt^{2}+\frac{a(t)^{2}}{[1+\frac{1}{4}kr^{2}]^{2}}(dr^{2}+r^{2}d\Omega^{2}),d\Omega^{2}\equiv d\vartheta^{2}+\sin^{2}\vartheta d\varphi^{2}$;
- perfect fluid (to assure isotropy + homogeneity): $\sigma_{ab}=\omega_{ab}=\dot{u}^{a}=0$;
- velocity field: $u^{0}=1,u^{i}=0$;

- expansion: $\theta(t) = 3\frac{\dot{a}(t)}{a(t)}$.
- EFE equations = Friedmann equations:

$$\frac{\ddot{a}}{a} = -\frac{1}{6}(\mu + 3p) + \frac{1}{3}\Lambda;$$
$$\mu = 3\frac{k}{a^2} + 3(\frac{\dot{a}}{a})^2 - \Lambda;$$
$$p = -2\frac{\ddot{a}}{a} - \frac{k}{a^2} - (\frac{\dot{a}}{a})^2 + \Lambda.$$

We introduce:

- the Hubble constant: $H \equiv \frac{1}{3}\theta = \frac{\dot{a}}{a}$,
- the deceleration parameter: $q \equiv -\frac{\ddot{a}a}{\dot{a}^2}$,
- redshift: $1+z \equiv \frac{\nu_{em}}{\nu_{obs}} = \frac{a(t=0)}{a(t)}$.
- the density parameters (under the inflation model as unitarian condition):

$$\Omega_m \equiv \frac{\mu}{3H^2} = -\frac{k}{H^2a^2} - \frac{\Lambda}{3H^2} + 1;$$
$$\Omega_k = -\frac{k}{H^2a^2}, \Omega_\Lambda = \frac{\Lambda}{3H^2};$$
$$\Omega_m + \Omega_k + \Omega_\Lambda = 1.$$

then the cosmological evolution (flow) of the Universe, during the BB evolution, is representing:

- as a sequence of evolutive epochs;

$$Q \rightarrow \underbrace{dS^- \rightarrow \text{inflation}} \rightarrow \underbrace{\text{radiation} \rightarrow \text{matter}} \rightarrow \underbrace{\text{present epoch}, \Lambda \rightarrow dS^+}$$

- epochs in which different kinds of fluids are dominant (scalar fields, Λ , radiation, mater and again Λ and possible scalar fields);
- epochs organized in three consecutive phases of acceleration, deceleration and again acceleration of the flow (the three outlined period in the above diagram).

The curvature parameter expresses the type of the universe we take into account: flat ($k = 0$), closed ($k > 0$) or open ($k < 0$).

PERTURBED FLRW COSMOLOGICAL FLOW

The study of the perturbed FLRW flow is related to the gauge problem, taking into account perturbations in FLRW metric: $g_{ab} = g^0_{ab} + \delta g_{ab}$,
or/and in the FLRW matter stress energy tensor: $\delta(x) = \frac{\mu(x) - \mu^0(x)}{\mu(x)}$.
We introduce the new variables: $X_a \equiv \tilde{\nabla}_a\mu, D_a \equiv a\frac{X_a}{\mu}, D \equiv (D_aD^a)^{1/2}, Z_a \equiv a\tilde{\nabla}_a\theta$.
Using these variables we obtain the grow of inhomogeneity of:
- first order:

$\dot{X}^{(a)} + \frac{4}{3}\theta X^a = -(\mu + p)Z^a - \sigma^a{}_b X^b + \eta^{abc}\omega_b X_c;$
$\dot{Z}^{(a)} + \theta Z^a = -\frac{1}{2}X^a - \sigma^a{}_b Z^b + \eta^{abc}\omega_b Z_c + \dot{u}^a \mathfrak{R} - 2\tilde{\nabla}^a(\sigma^2 - \omega^2) + \tilde{\nabla}^a(\tilde{\nabla}_b \dot{u}^b + \dot{u}_b \dot{u}^b);$
$\mathfrak{R} = -\frac{1}{3}\theta^2 + \tilde{\nabla}_a \dot{u}^a + (\dot{u}_a \dot{u}^a) - 2\sigma^2 + 2\omega^2 + \mu + \Lambda.$

- second order:

$$\ddot{D}_{(a)} + (\frac{2}{3} - 2w + c_s^2)\theta \dot{D}_{(a)} + [(\frac{1}{2} + 4w - \frac{3}{2}w^2 - 3c_s^2)\mu + \\ +(c_s^2 - w)\frac{12k}{a^2}]D_a + c_s^2[\frac{2k}{a^2}D_a - \tilde{\nabla}^2 D_a] = 0.$$

Perturbed flows could be followed in different evolutive epochs of the FLRW universe, and at different flow scales. In this case the universe becomes an 'almost FLRW universe' with different scales of inhomogeneities and anisotropies.

Early universe epochs

When we describe the cosmological flows in early BB universe we must first describe the exit from the pre-BB or the quantum gravitational (QG) phases, taking into account the possible mutual transformations: $(A)dS \leftrightarrow dS \leftrightarrow AsdS$.
But, the phase $dS^- \leftrightarrow AsdS$ is unstable, driven the excitation of the quantum fluctuations, in the form of a scalar field flow:

$\phi = \phi(t);$
$(\frac{\dot{a}}{a})^2 + \frac{k}{(a)^2} = \dot{\phi}^2 + V(\phi);$
$\frac{\ddot{a}}{a} = -(n-1)\dot{\phi}^2 + V(\phi);$
$\ddot{\phi} + n\frac{\dot{a}}{a} = -\frac{1}{2}\frac{\partial V}{\partial \phi}.$

where in the case of a Λ constant model we have: $\Lambda = \frac{n(n-1)}{2}V(\phi_0)$, and, $\gamma = 0, \mu = \Lambda$.
This form of the scalar field instability could also be used for the future epoch: $AsdS \rightarrow dS^+$.
During the inflationary phase, the instability flow, in the lower approximation, could be described by the equation: $\ddot{\phi} + 3H\dot{\phi} + V' = 0$.
Because of the inflation the general almost FLRW flow (known as 'Hubble flow') has a superhorizon scale.
We define the inflation parameters, EOS: $\varepsilon \equiv \frac{3}{2}(1 + w_\phi)$,
and, the deviation from a de Sitter expansion ($d^2\phi_0/dt^2 = 0$): $\delta \equiv \frac{\ddot{\phi}_0}{\dot{\phi}}(\frac{\dot{a}}{a})^{-1} - 1$.
We use the slow roll limit approximation: $\varepsilon, \delta \ll 1$. We then make the change of variable: $x \equiv a\phi$ and $d\eta^2 = a^2 dt^2, \tilde{\eta} = \eta - \eta_{endinfl}$. In this case the perturbation equation is simply: $\ddot{u} + [k^2 - 2(\frac{\dot{a}}{a})^2]u = 0$.
For $|k\tilde{\eta}| \gg 1$ (early times, inside Hubble horizon), the field behaves as a free oscillator: $\lim_{|k\tilde{\eta}| \to \infty} x = Ake^{-k\tilde{\eta}}$. The corresponding fluctuations could be quantized as: $\hat{x} = x(k,\eta)\hat{a} + x^*(k,\eta)\hat{a}^+$, and normalized to zero point fluctuation in the Minkowski vacuum: $x\,(k,\eta) = \frac{1}{\sqrt{2k}}e^{-k\tilde{\eta}}$.

For $|k\tilde{\eta}| \ll 1$ (late times, outside Hubble horizon), the fluctuations freeze in: $\lim_{|k\tilde{\eta}|\to\infty} x = iHaA, \phi = iHA$.

Cosmic Microwave Background (CMB)

The interpretation of the CMB anisotropies formation could be inferred using a photon gas: $p^a = E(u^a + e^a), E = (-p_a p^a), e_a e^a = 1, e_a u^a = 0$,
and an angular harmonic distribution: $f(\boldsymbol{x}, \boldsymbol{p}) = \sum_{l\geq 0} F_{A_l}(\boldsymbol{x}, E) e^{A_l}$.
Defining the brightness anisotropy multipoles: $\Pi_{A_l} = \int_0^\infty E^3 F_{A_l} dE$, the nonlinear *1+3* covariant multipole equations become:

$$
\begin{aligned}
&\dot{\mu}_R + \tfrac{4}{3}\theta\mu_R + \tilde{\nabla}_a q^a_R = 0;\\
&\dot{q}^{(a)}_R + \tfrac{4}{3}\theta q^a_R + \tfrac{4}{3}\mu_R \dot{u}^a + \tfrac{1}{3}\tilde{\nabla}^a \mu_R + \tilde{\nabla}_b \pi^{ab}_R = n_E\sigma_T(\tfrac{4}{3}\mu_R u^a_R - q^a_R);\\
&\dot{\pi}^{ab}_R + \tfrac{4}{3}\theta\pi^{ab}_R + \tfrac{8}{15}\mu_R\sigma^{ab}_R + \tfrac{2}{5}\tilde{\nabla}^{(a} q^{b)}_R + \tfrac{8\pi}{35}\tilde{\nabla}_c \Pi^{abc} = -\tfrac{9}{10} n_E\sigma_T\pi^{ab}_R;\\
&\dot{\Pi}^{(A_l)} + \tfrac{4}{3}\theta\Pi^{A_l} + \tilde{\nabla}^{(A_l}\Pi^{A_{l-1})} + \frac{(l+1)}{(2l+3)}\nabla_b\Pi^{A_l b} = -n_E\sigma_T\Pi^{A_l}.
\end{aligned}
$$

If we introduce the temperature fluctuation: $T(\boldsymbol{x}, \mathrm{e}) = T(\boldsymbol{x})\,[1 + \tau(\boldsymbol{x}, e)]$,
we obtain the corresponding equations for the temperature anisotropy multipoles:

$$
\begin{aligned}
&\frac{\dot{T}}{T} = -\tfrac{1}{3}\theta - \tfrac{1}{3}\tilde{\nabla}_a\tau^a;\\
&\dot{\tau}^{(a)} = -\frac{\tilde{\nabla}^a T}{T} - \dot{u}^a - \tfrac{2}{3}\tilde{\nabla}_b\tau^{ab} + n_E\sigma_T(v^a_R - \tau^a);\\
&\dot{\tau}^{(ab)} = -\sigma^{ab} - \tilde{\nabla}^{(a}\tau^{b)} - \tfrac{3}{7}\tilde{\nabla}_c\tau^{abc} - \tfrac{9}{10} n_E\sigma_T\tau^{ab};\\
&\dot{\tau}^{(A_l)} = -\tilde{\nabla}^{(A_l}\tau^{A_{l-1})} - \frac{(l+1)}{(2l+3)}\tilde{\nabla}_b\tau^{A_l b} - n_E\sigma_T\tau^{A_l}.
\end{aligned}
$$

Radiation epoch

In the case of pure radiation, $\gamma = \frac{4}{3}$, and the second order perturbation is obtained in the form: $\ddot{D}_{(a)} + \frac{1}{3}\theta\dot{D}_{(a)} - \frac{2}{3}(\frac{1}{3}\theta^2 + \frac{3k}{a^2})D_a + \frac{1}{3}[\frac{2k}{a^2}D_a - \tilde{\nabla}^2 D_a] = 0$.

Matter epoch

During the matter dominance epoch we can obtain the equations for a cold dark matter (collisionless gas, CDM) modeled as a 'dust fluid', and for an ordinary 'fluid with collisions' (i.e. baryonic matter) .
In the case of dust, $p = 0$, $\dot{u}^a = 0$ and the second order perturbation is:
$\ddot{D}_{(a)} + \frac{2}{3}\dot{D}_{(a)} - \frac{1}{2}\mu D_a = 0$.
The solution of this equation has the form: $D_a(t, x^\alpha) = D_{+a}(x^\alpha)t^{2/3} + D_{-a}(x^\alpha)t^{-1}$, where *t* represents the proper time along the flow lines. This shows the growing mode that leads to CDM structure formation and the decaying mode that dissipates previously existing inhomogeneities.

The case of a fluid with collisions is manifest between - 4/3 $\leq w \leq 2$.
The two limits represent, respectively, the case of a 'stiff matter' (where the speed of sound equals the speed of light) and the case of a 'matter with no active gravitational mass'. The full equation of the second order perturbation must be used in this case. The instabilities that appear in the flow are related to the Jeans' criterion of gravitational collapse:

$$\frac{1}{2}(1-w)(1+3w)\mu D_a > w[\frac{2k}{a^2}D_a - \tilde{\nabla}^2 D_a].$$

The instabilities that appears could have different scales:

- Jeans length: $\lambda_J = c_s\sqrt{\frac{\pi}{G\mu(1-w)(1+3w)}}$;
- short-wavelength solutions: $\lambda < \lambda_J$;
- long-wavelength solutions: $\lambda > \lambda_J$.

Until now, starting from an assigned Cosmological Principle, we have discussed perturbed 'almost FLRW' cosmologies, with a defined connection at the CMB epoch. Inversely, taking into account the cosmological observations, we have the following theorem and corrolaries:
Theorem (Ehlers, Geren, Sachs [2]): If the fundamental observers in a dust spacetime see an isotropic radiation field, the spacetime is locally (almost) FLRW.
Corollary 1 An irrotational perfect fluid spacetime that admits an isotropic radiation field has spherical, planar or hyperbolic symmetry and admits a strict thermodynamic scheme.
Corollary 2 Any solution to EFE in which the matter non-interacting radiation, expanding dust (CDM), and scalar field (or cosmological constant), and for which the dust sees an isotropic radiation field, must be either an FLRW model, or have the gradient of the scalar field orthogonal to the dust congruence.
These theoretical results, together with the CMB observations, which impose a limiting values to the perturbations of $\delta\mu/\mu < 10^{-5}$, allow us to consider the congruence at CMB, and to attribute to the mathematical spacetime, physical properties and thermodynamics.

Backreaction of the perturbations on the FLRW flow

Having chosen a foliation of spacetime, we obtain two 'directions' of the flow. One will be extrinsic (on time), in the direction of the extrinsic curvature $K_{ab} \equiv \theta_{ab}$.
The other is intrinsic (on the three dimensional spatial hypersurfaces), in the direction of the Ricci tensor R_{ab}. So, we have the evolutionary equations:
$K_{ab} = -\frac{1}{2}\partial_t g_{ab}, R_{ab} = -\frac{1}{2}\frac{\partial}{\partial r}g_{ab}$.
Taking into account different perturbations and inhomogeneities, we can define the spatial averaging of a scalar field Ψ on a spatial compact domain Δ , through its Riemannian volume average: $\langle\Psi(X^i,t)\rangle \equiv \frac{1}{V_\Delta}\int_\Delta \Psi(X^i,t)\sqrt{\det(g_{ab})}d^3X$.

We can rewrite the FLRW equations in the form:

$$3(\frac{\dot{a}_\Delta}{a})^2 - 8\pi G\langle\rho\rangle_\Delta - \Lambda = -\frac{\langle R\rangle_\Delta + Q_\Delta}{2};$$
$$3\frac{\ddot{a}_\Delta}{a_\Delta} + 4\pi G\langle\rho\rangle_\Delta - \Lambda = Q_\Delta.$$

where the kinematical back-reaction arises as a result of expansion and shear fluctuations: $Q_\Delta \equiv 2\langle II\rangle_\Delta - \frac{2}{3}\langle I\rangle_\Delta^2 = \frac{2}{3}\langle(\theta - \langle\theta\rangle_\Delta)^2\rangle_\Delta - 2\langle\sigma^2\rangle_\Delta$.
where I and II denote the principal scalar invariants of the extrinsic curvature. The back-reaction term and the average Ricci tensor act as an additional scalar field with EOS:

$$\rho_{eff}^\Delta \equiv \langle\rho\rangle_\Delta + \rho_\phi^\Delta, p_{eff}^\Delta \equiv p_\phi^\Delta;$$
$$\rho_\phi^\Delta = -\frac{1}{16\pi G}Q_\Delta - \frac{1}{16\pi G}R_\Delta, p_\phi^\Delta = -\frac{1}{16\pi G}Q_\Delta - \frac{1}{48\pi G}R_\Delta.$$

in a matter FLRW universe. This could be inferred as an explanation for the present re-acceleration phase of the matter universe in the nonlinear phase of the growth of the perturbations .

SIMULATIONS OF THE COSMOLOGICAL FLOWS

As we discussed in the last paragraph, we can model the cosmological flows on different scales, from global ones (cosmological, at the level and even upper the Hubble horizon, $d \gg 1000h^{-1}$ Mpc), to local ones (cosmic, at a level $d \leq 100h^{-1}$ Mpc).
In our simulations we are interested in the cosmic simulations of the evolution of the local universe - on both 'large scale universe' (LSS, $d \sim 100h^{-1}$ Mpc) and intermediate scales ($d \sim 25-50h^{-1}$ Mpc) - to take into account the formation and the evolution of the CDM+baryonic matter components (pancakes, filaments, galactic and supergalactic components, voids, filamentary second order WEB structures).
As fluids we take into account a dust component (CDM) and a collisional one (baryonic), in both linear and nonlinear evolutive phases ($100 > z > 0$), with an additional extra Λ to take into account the recent accelerated phase of the evolution (as result, for ex., of an extra backreaction field).
We study different additional phenomena in the matter components (virialized/unvirialized, dense/underdense) and additional heating/cooling mechanisms.
In respect to the density level of the perturbations we can define different kinds of matter components: galactic halos (ultra-dense matter, $\rho \geq 200\langle\rho\rangle$), pancakes and filaments (dense matter $5\langle\rho\rangle < \rho < 200\langle\rho\rangle$) and voids (under-dense regions, $\rho \leq 5\langle\rho\rangle$).
The evolutionary equations for both collisionless and collisional fluids can be obtained from the scalar components of the gauge transformations.
To study the collisionless fluid we introduce the co-moving variables: $\mathbf{x} = \mathbf{x}\,(t)$, $\mathbf{r} = a(t)\mathbf{x}$, $\mathbf{p} = ma^2\partial_t\mathbf{x}$.

The evolutive equations are:

$$\frac{d\mathbf{x}}{da} = \frac{\mathbf{p}}{\dot{a}a^2};$$
$$\frac{d\mathbf{p}}{da} = -\frac{\nabla\phi}{\dot{a}};$$

$$\nabla^2\phi = 4\pi G\Omega_m(t)a^2\rho_{cr}(t)\delta.$$

To describe the collisional fluid it is better to use a Lagrangian scheme with a smoothing length scale. We introduce the Lagrangian variables: $x^i(\mathbf{q},\tau) = q^i + \psi(\mathbf{q},\tau), \rho(\mathbf{q},\tau) = \langle\rho\rangle|\frac{\partial x^i}{\partial q^i}|, \frac{d}{d\tau} \equiv \frac{\partial}{\partial\tau} + v^j\nabla_j$.
The set of equations, in Lagrangian formalism for the collisional fluid are:

$$\frac{d\delta}{d\tau} + (1+\delta)\theta = 0;$$
$$\nabla_i v_j = \frac{1}{3}\theta\delta_{ij} + \sigma_{ij} + \omega_{ij};$$
$$\frac{d\theta}{d\tau} + \frac{\dot{a}}{a} + \frac{1}{3}\theta^2 + \sigma^{ij}\sigma_{ij} - 2\omega^2 = -4\pi Ga^2\langle\rho\rangle\delta;$$
$$\frac{d\omega^i}{d\tau} = \frac{\dot{a}}{a}\omega^i + \frac{2}{3}\theta\omega^i - \sigma^i{}_j\omega_j = 0;$$
$$\frac{d\sigma_{ij}}{d\tau} + \frac{\dot{a}}{a}\sigma_{ij} + \frac{2}{3}\theta\sigma_{ij} + \sigma_{ik}\sigma^k{}_j + \omega_i\omega_j - \frac{1}{3}\delta_{ij}(\sigma^{kl}\sigma_{kl} + \omega^2) = -E_{ij};$$
$$\frac{dE_{ij}}{d\tau} + \frac{\dot{a}}{a}E_{ij} - \nabla_k\varepsilon^{kl}{}_{(i}H_{j)l} + \theta_E ij + \delta_{ij}\sigma^{kl}E_{kl} - 3\sigma^k{}_{(i}E_{j)k} = -4\pi Ga^2\rho\sigma_{ij};$$
$$E_{ij} \equiv T_{ij} = \nabla_i\nabla_j - \frac{1}{3}\delta_{ij}\nabla^2\phi;$$
$$H_{ij} \equiv -\frac{1}{2}\nabla_{(i}H_{j)} - 2v_k\varepsilon^{kl}{}_{(i}E_{j)l}.$$

For a continuous flow, we used smoothed values estimated over a scale h: $A_s(\mathbf{x},t) \equiv \int d^3\mathbf{x}' A(\mathbf{x}',t)W(\mathbf{x}-\mathbf{x}',\mathbf{h})$.
The additional heating (Q) and cooling phenomena introduce an extra term in the energy equation in the form: $\propto \frac{Q-L}{\rho}$.
In this mode we can study:
Global cosmical flow:

- the peculiar velocity field at large scale has a Hubble type characteristic; linear in position; with super Hubble form: $\dot{\delta} + \frac{1}{a}\nabla\mathbf{v} = 0; \partial_t\mathbf{v} + \frac{\dot{a}}{a}\mathbf{v} = -\frac{\nabla\phi}{a} \rightarrow \mathbf{v} = \frac{2}{3}\frac{f}{H\Omega} - \frac{\nabla\phi}{a} + \frac{\nabla\times U(\mathbf{x})}{a}$;

- non-linear correction of the flow:

$$\delta(\mathbf{x},t) = \delta^1(\mathbf{x},t) + \delta^2(\mathbf{x},t) + \ldots \delta^2(\mathbf{x},t) = D^2(t)[\frac{2}{3}(1+K(t)\varepsilon^2(\mathbf{x}) + \nabla\varepsilon(\mathbf{x})\cdot\nabla d + (\frac{1}{2} - k(t)T^2].$$

- cosmic structures - Gaussian by descendent, autocorrelation functions in the form: $\xi(\mathbf{r}) = \int \frac{d\mathbf{k}}{(2\pi)^3}P_f(k)e^{-i\mathbf{k}\cdot\mathbf{r}}$.

Voids:

- expand as over-dense regions collapse;
- spherical shape: slight asphericities decrease as the voids become larger.

The study of the **external tidal fields** is made in respect to the alignment of the tidal (T) and inertial (I) tensors:

- galactic spin: $\mathbf{M} = \int_V d^3 q \rho a^3 (a\mathbf{x} - a\bar{\mathbf{x}}) \times \frac{\mathbf{v}}{a}, \bar{\mathbf{x}} = \int_V d^3 q \frac{\mathbf{x}}{\mathbf{V}}$;
- initial angular momentum: $M_i = -a\dot{D}\varepsilon_{ijk} T_{jl} I_{lk}, T_{ij} = \nabla_i \nabla_j \Phi_i - \frac{1}{3}\delta_{ij}\nabla^2\Phi_i, I_{ij} = \int_V (q_l - \bar{q}_l)(q_k - \bar{q}_k)\rho_b a^3 d^3 q$.

For the **pancakes** and **halos** we can define the parameters:

- the deformation tensor D: $J = |\partial\mathbf{x}/\partial\mathbf{q}| = |\det D|$,
- the density contrast: $1 + \delta = \frac{1}{J}$,
- local values of the tidal field ($\lambda_1 \geq \lambda_2 \geq \lambda_3$): $1 + \delta = \frac{1}{[1-b(\tau)\lambda_1][1-b(\tau)\lambda_2][1-b(\tau)\lambda_3]}$,
- initial top hat model for the collapse of halos:

$$\delta = \begin{cases} \left|\frac{(\delta_c - \delta_i)/\bar{\delta}_c}{F(\Omega)(\cosh\theta - 1)}\right|^3 & \delta_i < \delta_c \\ \left|\frac{(\delta_c - \delta_i)/\bar{\delta}_c}{F(\Omega)(1 - \cos\theta)}\right|^3 & \delta_i \geq \delta_c, \end{cases}$$

where: $\delta_c = \frac{3}{5}(\frac{1}{\Omega} - 1), \bar{\delta}_c = \begin{cases} \delta_c & \Omega \neq 1 \\ 1 & \Omega = 1. \end{cases}$

- halos virial theorem:

$$\frac{1}{2}\frac{D^2 I}{D\tau^2} = 2 \cdot T + N + 3 \cdot (\gamma - 1) \cdot E - \int p \cdot \mathbf{r} \cdot d\mathbf{S},$$

where: $N = -(1/2)\, G \int_V \frac{\rho(\mathbf{x})\rho(\mathbf{x}')}{|\mathbf{x}-\mathbf{x}'|} d\mathbf{x} d\mathbf{x}', T = \frac{1}{2}\int \rho \mathbf{v}^2 d\mathbf{x}, I = \int \rho \mathbf{x}^2 d\mathbf{x}, E = \int \varepsilon d\mathbf{x}$.

- halos theory:

$$r_{200} = \left[\frac{GM}{1000\Omega_{DM}(z)H^2(z)}\right]^{1/3}, t_{cool}(r_{cool}) = H(z)^{-1}, V_c = \sqrt{\frac{GM}{r_{200}}}, T_{vir} = \frac{\mu m_p V_c^2}{2k},$$

- halos local analysis:
  - virialization parameter: $\beta = \frac{3(\gamma-1)E + 2T}{E - N}$;
  - turbulent Mach number: $Mach_{turb} = \frac{\mathrm{V}_{rms}}{C_s}$;
  - Rayleigh inviscid instability (related to the specific angular momentum j of the halo: $dj^2/dr^2 > 0$.

Some results of our cosmological simulations obtained in the Bucharest Observatory, using an ALTIX/SGI platform, were presented recently ([3]).

REFERENCES

1. R. Marteens, *Phys. Rev. D* **55**, 463–500 (1997).
2. J. Ehlers, P. Geren, R. K. Sachs, *J. Math. Phys.* **9**, 1344–1350 (1968).
3. M. D. Suran, "Cosmological Simulations in Bucharest Observatory", in *Fifty Years of Romanian Astrophysics, 2006*, edited by C. Dumitrache et al., AIP Conference Proceedings 895, American Institute of Physics, New York, 2007, pp. 283–287.

Interactions effects on color and morphology of galaxy pairs

Nedelia A. Popescu

Astronomical Institute of the Romanian Academy, Bucharest, Romania,
e-mail:nedelia@aira.astro.ro

Abstract. Properties of galaxy pairs, such as color and morphology, are analyzed with respect to the samples of galaxies in the field of the radio galaxies 3C 220.1 ($z = 0.62$) and 3C 34 ($z = 0.689$), covering a total field of 12.4 arcmin^2. Many galaxies present evident signs of interactions, with features of morphological disturbance and bluer optical-NIR colors, suggesting that close interactions and mergers induce star-forming episodes. The presence of many elliptical galaxies interacting in *dry mergers* is also distinguished in both samples of galaxy pairs.

Keywords: Galaxy interactions, colors and morphology of paired galaxies
PACS: 98.65.At, 98.65.Fz

INTRODUCTION

This paper intends to perform a complex and extended study on the properties of galaxy pairs with respect to their parent sample, as well as on the interaction effects on color and morphology of galaxy pairs in the field of the radio galaxies 3C 220.1 ($z = 0.62$, 64 galaxies)and 3C 34 ($z = 0.689$, 89 galaxies).

We have used a sample of ~ 150 galaxies with photometrical and morphological data (optical-NIR photometrical data (V, I, J, H, K) and HST/WFPC2 morphological data of Stanford et al. 2002 [1]), covering a total field of 2×6.2 arcmin^2.

By means of Z-PEG software (Le Borgne & Rocca-Volmerange, 2002 [2]) we determined the redshifts for galaxies in the field of the radio galaxies and we analyzed the clustering tendencies of galaxies with similar redshifts (for details see [3]). In our determinations we have used the cosmological parameters $H_0 = 65$ km/s/Mpc and $q_0 = 0.15$.

IDENTIFICATION OF CLOSE GALAXY PAIRS AND STRONGLY DISTURBED GALAXIES

Galaxy pairs statistics, like the merger rate itself, depend on clustering and the mean number density of galaxies in the sample. An integrated measure of galaxy clustering on small scales is provided by the close pairs statistics.

The analysis of the distribution of galaxy-galaxy separations in the morphological and photometric redshifts data of the studied catalogues of galaxies is performed as a quantitative measure of the excess of close pairs. The projected physical separation and rest-frame line-of-sight velocity difference can uniquely specify the pairs of galaxies in

CP934, *Flows, Boundaries, Interactions*
edited by C. Dumitrache, V. Mioc, and N. A. Popescu

the redshift space. The following conditions are usually considered:

$$10h^{-1}kpc < \triangle r_{proj} < r_{max}; \quad r_{max} = 30, 50, 100h^{-1}kpc \tag{1}$$

$$\triangle v < 500 km/s \tag{2}$$

The values $r_{max} = 100$ h^{-1} kpc and $\triangle v < 500$ km/s are adequate to define pairs with enhanced star formation activity. At the value $r_{max} = 30$ h^{-1} kpc, the effects of close interactions and authentic merger pairs can be determined. The definitions for close dynamical pairs are: 5h^{-1} kpc $< \triangle r_{proj} <$ 20h^{-1} kpc; $\triangle v < 500$ km/s (i.e. for a close companion that will probably merge soon, within 0.5 Gyr).

At redshifts of $z \sim 0.8 - 0.9$, the offsets between galaxies in the range $2'' - 10''$ correspond to projected distances of approximately 16-80 h^{-1} kpc. At these distances the systems could not be in the late stages of mergers, representing only bound systems that influence each other or systems affected by the same global environmental influences.

In our study we selected the galaxy pairs adopting the following thresholds:

a) projected relative distance: $\triangle r_{proj} < r_{max} = 100$ h^{-1} kpc ;

b) relative velocity: $\triangle v < 500$ km/s;

c) relative magnitude: $\triangle K < 1.75$ mag.

The 2D distributions of galaxies and galaxy pairs, function of HST/WFPC2 morphology and photometric redshifts, are presented in Figures 1 and 2. The symbols are as follows: squares - E/S0; circles - S; diamonds - Irr, mergers; (+) - galaxies outside the WFPC2 field. The galaxies with photometric redshifts in different ranges are overlapped on the mentioned symbols.

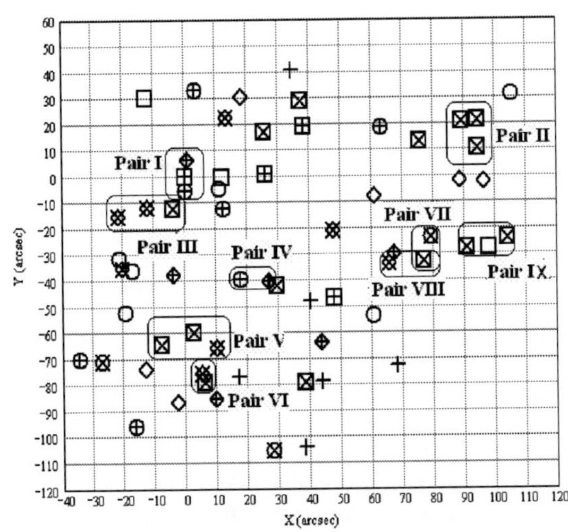

FIGURE 1. Spatial distribution of galaxies and galaxy pairs, function of morphology, in the field of 3C 220.1

For 3C 220.1 field we obtained 17 galaxies in the redshift range $0.45 < z < 0.7$ (denoted with (+)), 11 galaxies being located in the inner region of the field close to the brightest cluster galaxy BCG (an E/S0 galaxy with $K = 15.76$, $(I - K) = 2.74$,

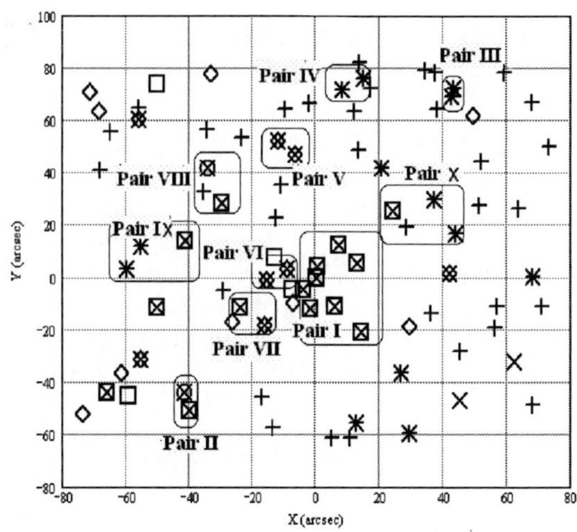

FIGURE 2. Spatial distribution of galaxies and galaxy pairs, function of morphology, in the field of 3C 34

$(J-K) = 1.7$). In the redshift range $0.75 < z < 1.05$, 26 galaxies marked with ($\times$), are clustered in three groups, containing 15 E/S0 galaxies.

In this field we determined 21 galaxies forming 9 galaxy pairs and triplets. Pairs I and IV consist of galaxies with redshifts in the range $0.45 < z < 0.7$ and projected distances smaller than 50 kpc.

The interaction is obvious in the case of Pair I. Because the BCG (E/S0) interacts with an irregular galaxy and a spiral one, this galaxy contributes to two separate bound pairs, forming a triplet system.

Pairs II, III, V-IX consist of galaxies with redshifts in the range $0.75 < z < 1.05$ and have mixed morphology (i.e. an elliptical/lenticular galaxy + a spiral or irregular galaxy). Pairs VII, VIII and IX form a filamentary structure (see Figure 1).

For 3C 34 field, the ($\times$) symbols represent 40 galaxies with photometric redshifts $0.65 < z < 1$. In the inner region of the field, close to BCG (an E/S0 galaxy with $K = 15.54$, $(I-K) = 2.97$, $(J-K) = 1.96$), 30 galaxies with $0.65 < z < 1$ are clustered.

In this field we determined 18 pairs containing 29 galaxies. Group I consists of 8 galaxies with redshifts in the range $0.65 < z < 1$ and projected distances smaller than 50 kpc. This region is the most interesting because the BCG (E/S0) interacts especially with other E/S0 galaxies. These sort of interactions are called *dry mergers* (i.e. mergers between non-star forming early-type galaxies, E/S0).

Pairs II-X consist of galaxies with redshifts in the range $0.65 < z < 1$ and have mixed morphology (see Figure 2).

COLOR-COLOR DISTRIBUTION OF GALAXIES FUNCTION OF MORPHOLOGY

In Figures 3 and 4 are presented the 2D and 3D color-color diagrams for the galaxies in the studied samples.

It is obvious that E/S0 galaxies are located in the color-color plane around $5.5 < (V-K) < 6$, $0.8 < (J-H) < 1$ and $0.6 < (H-K) < 0.8$, having colors corresponding to the redshifts $z \sim 0.7-0.9$ (as their photometric redshifts obtained via Z-PEG model).

These E/S0 galaxies represent the components of possible groups or clusters of galaxies at the redshifts $z \sim 0.7-0.9$, as it can also be observed from the photometric redshifts distribution of the galaxies in the field of 3C 220.1 and 3C 34 (for details see [4]).

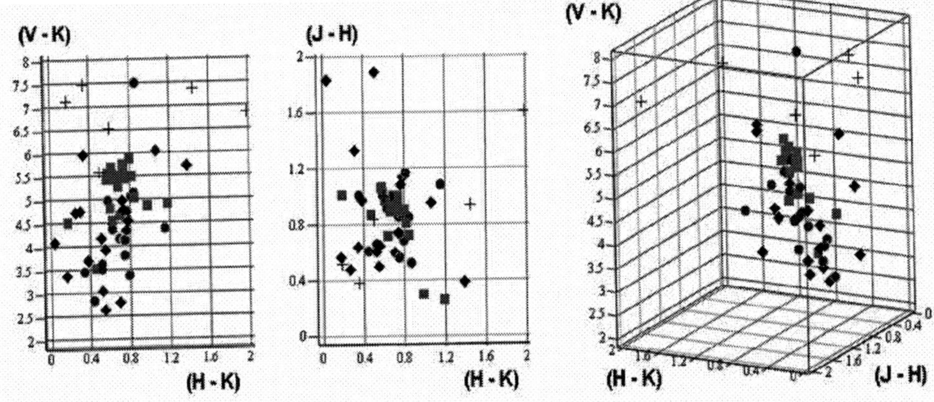

FIGURE 3. 3C 220.1 field: 2D (left and middle) and 3D (right) $(V-K)$, $(J-H)$ and $(H-K)$ colors distribution of galaxies function of morphology; squares - 22 E/S0 gal.; circles - 16 S gal.; diamonds - 20 Irr, mergers gal.; (+) - 6 galaxies outside the WFPC2 field

CLOSE GALAXY PAIRS COUNTING

The relative density δ is obtained by binning the sample of galaxy into rings defined by $R_{\max}$ and $R_{\min}$, and summing over all galaxies for a given galaxy-galaxy separation R_{sep}:

$$\delta = \frac{\sum_i^j (N^i)}{\pi(R_{\max}^2 - R_{\min}^2)} \tag{3}$$

where j is the total number of galaxies in our sample.

If the galaxies are uniformly distributed, the distribution of δ function of R_{sep} will be flat.

In order to determine whether interactions and mergers are more prevalent in the studied fields, we investigated the distribution of galaxy-galaxy separations.

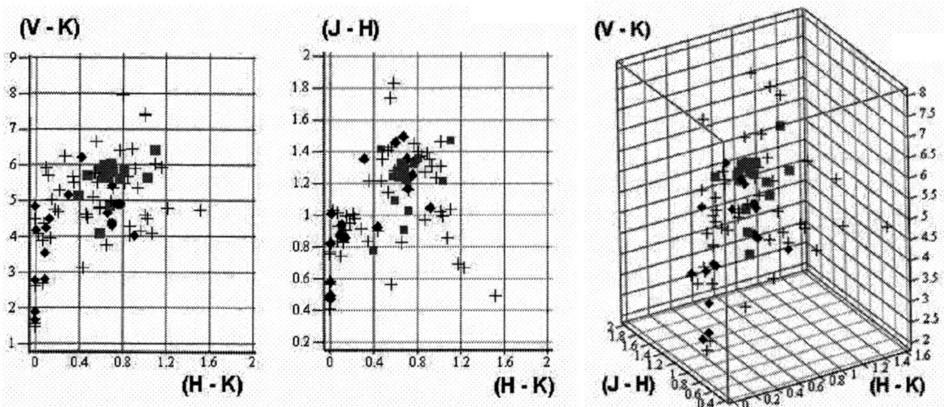

FIGURE 4. 3C 34 field: 2D (left and middle) and 3D (right) $(V-K)$, $(J-H)$ and $(H-K)$ colors distribution of galaxies function of morphology; squares - 19 E/S0 gal.; circles - 2 S gal.; diamonds - 17 Irr, mergers gal.; (+) - 51 galaxies outside the WFPC2 field

Using the above formula we determined the relative galaxy density around 22 red E/S0 galaxies, in the field of 3C 220.1, and around 19 E/S0 galaxies in the 3C 34 field (see Figure 5).

The number of galaxies in each bin ($5''$) is weighted by the area. A flat distribution would show that galaxies in the outer parts of the clusters are uniformly distributed.

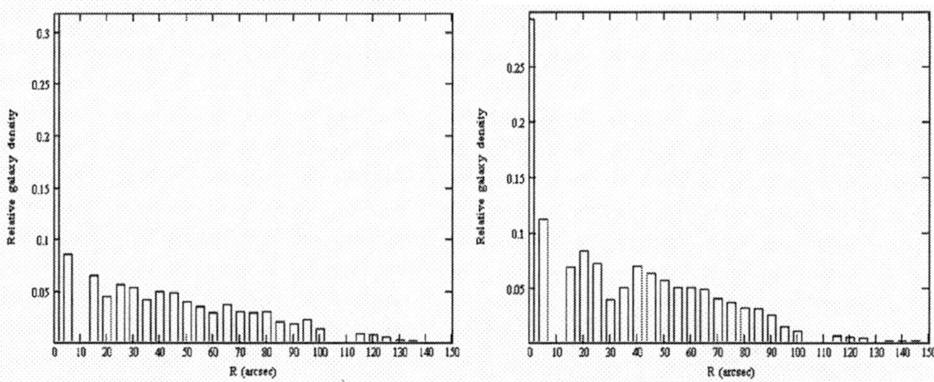

FIGURE 5. The relative galaxy density for 3C 220.1 (left) and 3C 34 (right) fields

The relative galaxy density presents an excess of pairs at small separations ($10''$) in both fields.

For the relative galaxy distribution in 3C 34 field (Figure 5) an excess of galaxy pairs at separations of $20''$ and $40''$ can be observed too.

PROPERTIES OF GALAXY PAIRS WITH RESPECT TO THEIR PARENT SAMPLE

In our study, at least half of the pairs exhibit clear morphological signs of on-going interactions.

Figure 6 presents the distribution of morphologies in the galaxy pairs (solid histogram) and parent samples (dotted histogram).

The morphology type is as follows: 0 - galaxies without a determined type; 1 - E/S0; 2 - Sa/Sb; 3 - Sc/Sd; 4 - Irr; 5 - mergers/interactions.

Comparison of the morphologies shows that there are marginally more E/S0 galaxies and Scd/Irr galaxies among pairs (for 3C 220.1 field), while there are fewer Sab galaxies.

The increased fraction of E/S0 type galaxies in pairs may be a result of early merging in groups. We mention the presence of a great number of *dry mergers*: 8 E/S0 galaxies in Group I of 3C 34 field and 7 E/S0 galaxies in 3C 220.1 field. We also observed the presence of dry mergers at projected galaxy-galaxy separation less than 20 kpc, relative magnitude $K < 1.75$ mag and photometric redshifts difference $\Delta z < 0.1$.

Also, the excess of very late-type galaxies can be explained by the presence of interactions that increase the star formation rate, leading to later classification.

The solid and dotted histograms in Figures 7 and 8 represent the distribution in colors of galaxy pairs and parent samples, respectively, for 3C 220.1 and 3C 34 fields.

Galaxies that present clear features of interactions, with signs of morphological disturbance and bluer colors, suggest that close interactions and mergers induce star formation episodes.

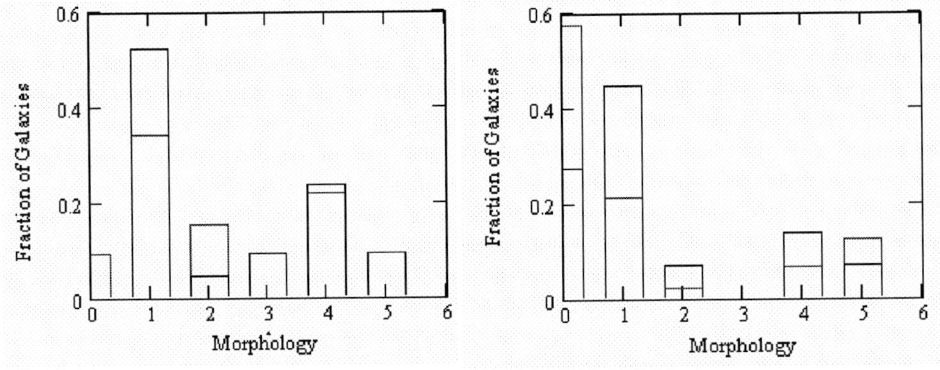

FIGURE 6. The distribution of morphologies of galaxy pairs and parent samples for 3C 220.1 (left) and 3C 34 (right) fields

CONCLUSIONS

In our study we have combined the photometrical and morphological data for ~150 galaxies in the dense environments of 3C 220.1 and 3C 34 radio galaxies, in order to analyze the frequency of galaxy close pairs, the properties of galaxy pairs with respect

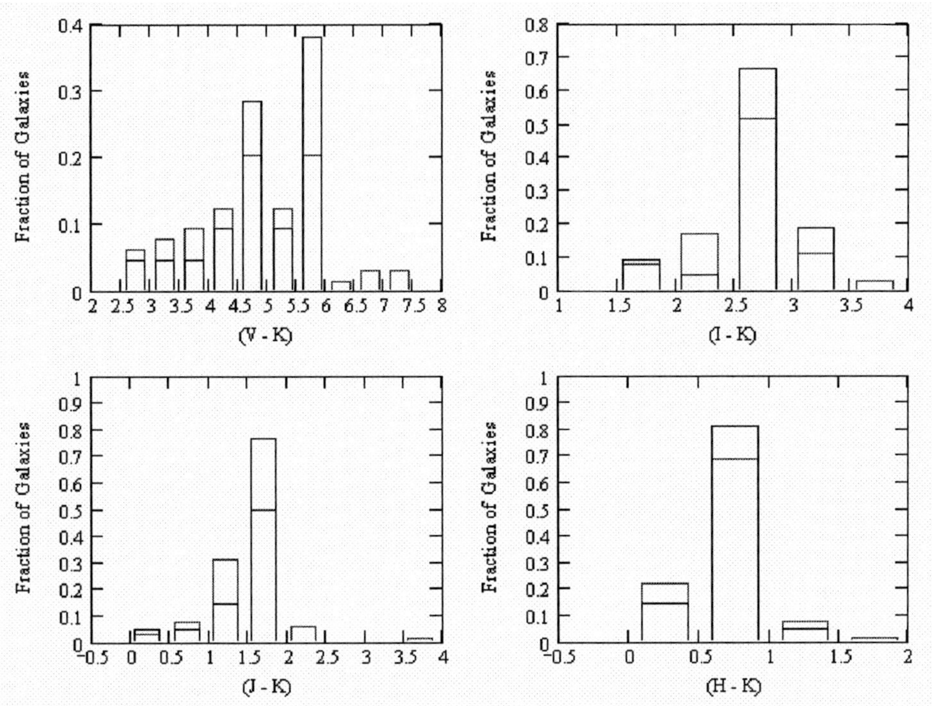

FIGURE 7. The distribution in colors of galaxy pairs and parent samples of 3C 220.1

to their parent sample and the interaction effects on color and morphology of galaxy pairs.

A complete sample of galaxy pairs has been obtained considering the thresholds on projected relative distance ($r_{max} = 100\ \mathrm{h}^{-1}$ kpc), relative velocity ($\triangle v < 500$ km/s) and relative magnitude ($\triangle K < 1.75$ mag).

In the samples of 9 galaxy pairs for 3C 220.1 field and 18 galaxy pairs for 3C 34 field, we determined 15 galaxies involved in *dry mergers* and 17 mixed pairs, at least half of the pairs exhibiting clear morphological signs of on-going interactions.

Elliptical galaxies in E+S pair or E+Irr pair present bluer $J-H$ and $V-K$ colors than expected, suggesting that close interactions and mergers induce star formation episods.

The influence of the environment on both galaxies with bright companions and faint companions seems to be comparable.

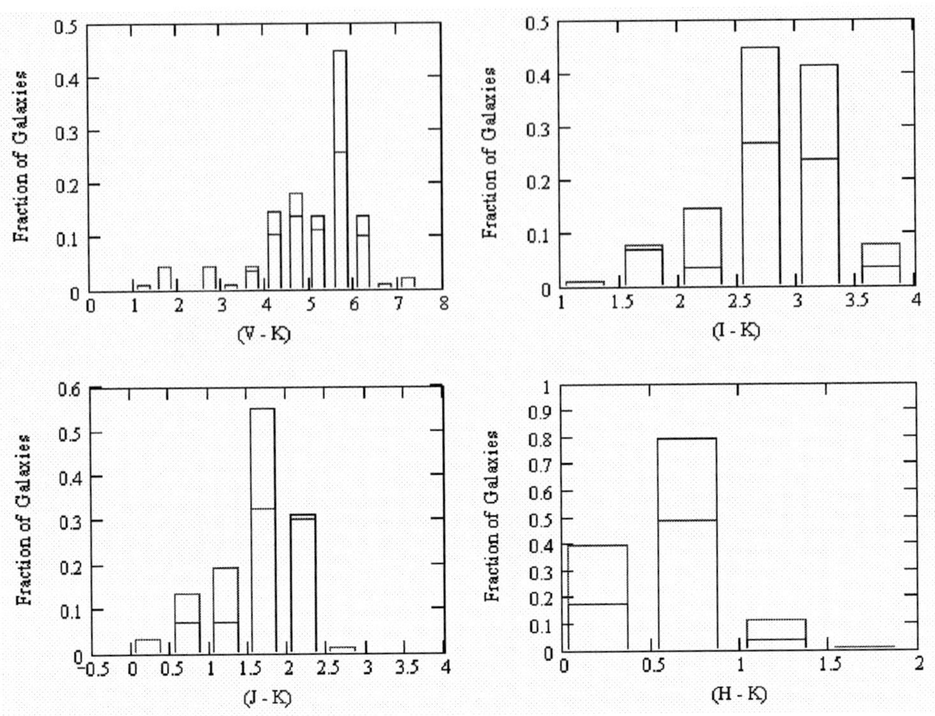

FIGURE 8. The distribution in colors of galaxy pairs and parent samples of 3C 34

REFERENCES

1. S. A. Stanford, P. R. Eisenhardt, M. Dickinson, B. P. Holden and R. De Propris, *Astrophys. J. Suppl.*, **142**, 153-160 (2002).
2. D. Le Borgne and B. Rocca-Volmerange, *Astron.Astrophys*, **386**, 446-455 (2002).
3. N. A. Popescu, *Fifty Years of Romanian Astrophysics, AIP Conference Proceedings*, eds. C. Dumitrache, N.A. Popescu, M.D. Suran, V. Mioc, **895**, 341-345 (2007).
4. N. A. Popescu, *Fifty Years of Romanian Astrophysics, AIP Conference Proceedings*, eds. C. Dumitrache, N.A. Popescu, M.D. Suran, V. Mioc, **895**, 294-299 (2007).

The Photometry of Poorly Studied Open Star Clusters in the Milky Way

Zabolotskikh M.V.[*], Glushkova E.V.[†,*], Koposov S.E.[**,*], Spiridonova O.I.[‡] and Rastorguev A.S.[†,*]

[*]*Sternberg Astronomical Institute, Moscow University, 13 Universitetskii prospect, Moscow, 119991, Russia*
[†]*Faculty of Physics, Moscow University, 1, bld. 2 Leninskie Gory, Moscow, 119992, Russia*
[**]*Max Planck Institut fur Astronomie, 17 Konigstuhl, Heidelberg, D-69117, Germany*
[‡]*Special Astrophysical observatory, Nizny Arkhyz, 369167, Karachai-Circassian Republic, Russia*

Abstract. BVRI CCD photometric survey of poorly studied galactic open star clusters in the direction to Perseus spiral arm was performed with 1-meter Zeiss telescope of SAO RAS. First determinations of basic parameters for King 18 star clusters derived from original CCD BVR_CI_C observations are presented and compared with NIR data. The cluster has been identified as very young, with an age $\sim$ 30 Myr, locating at the distance $\sim$ 2.9 kpc.

Keywords: open star clusters, distance scale, galactic structure and kinematics, stellar evolution
PACS: 97.10.Vm, 98.20.-d, 98.20.Di, 98.35.Ac, 98.35.Hj, 98.35.Ln

INTRODUCTION

Open star clusters remain among most important and interesting objects in our Galaxy. The interest is constantly supported by their unique role in wide fields of modern astronomy, from stellar evolution to the structure and dynamics of the Galaxy as a whole. They are most suitable objects for age determination and the study of star-formation history. Young open clusters are also the tracers of stellar complexes and galactic spiral pattern. Isochrone fitting technique enables distance determination, and even recent Cepheid distance scale is based mainly on the distances of Cepheids – members of young open clusters. Small internal velocity dispersion of cluster members in many cases provides us with accurate radial velocity and proper motion determinations, giving rise to kinematic study of the galactic disk.

The total number of known open clusters approaches to two thousands, but there is a substantial lack of modern photometric data for more than one thousand clusters, which makes impossible the estimates of their basic parameters, such as distances and ages. New discoveries of large population of open clusters [1-4] in "all-sky" infrared catalogs, hidden by interstellar absorption in optics, encourage our efforts to carry out the photometric investigation of poorly studied open star clusters.

CP934, *Flows, Boundaries, Interactions*
edited by C. Dumitrache, V. Mioc, and N. A. Popescu

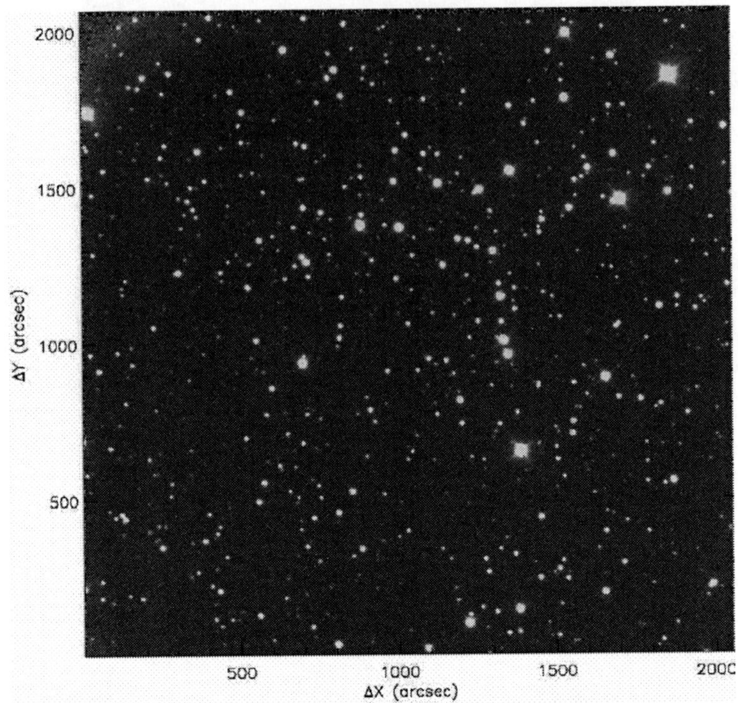

FIGURE 1. Cluster image in V band. Faintest stars have $21^m(V)$.

THE PROGRAM

At first, we have selected clusters in large segment of Perseus spiral arm. It is well known, that its extension to the third galactic quadrant is deficient with young population, so it is hardly seen in all recent star surveys. In the future prospect this cluster sample promises the way to sketch the spiral arm cross-section and, in combination with radial velocity measurements, to measure galactic rotation speed far from the galactic center, where recent rotation curves seem to be unreliable.

The list of poorly studied clusters in Perseus arm, observed by our team in 2002-2004, includes 30 objects: Berkeley 5, 6, 43, 45, 94 and 95; IC 166 and 1442; King 1, 4, 9, 12, 13, 14, 16, 18, 19, 20, 21, 25 and 26; NGC 266, 609, 2254, 6802, 7226, 7245, 7261, 7296 and 7788. In this paper we demonstrate preliminary results of the photometric study of King 18 open cluster in BVR_CI_CJHK bands. Optical CCD data for this cluster have been obtained for the first time.

The cluster image in V band is shown in fig. 1, where some excess of relatively bright stars can be seen. The photometrical limit estimates approximately as $21^m(V)$ for five minutes exposure. Its equatorial coordinates are $\alpha_{J2000} = 22^h52.1^m, \delta_{J2000} = +58°17'$; galactic coordinates $(l = 107.8°, b = -1.0°)$.

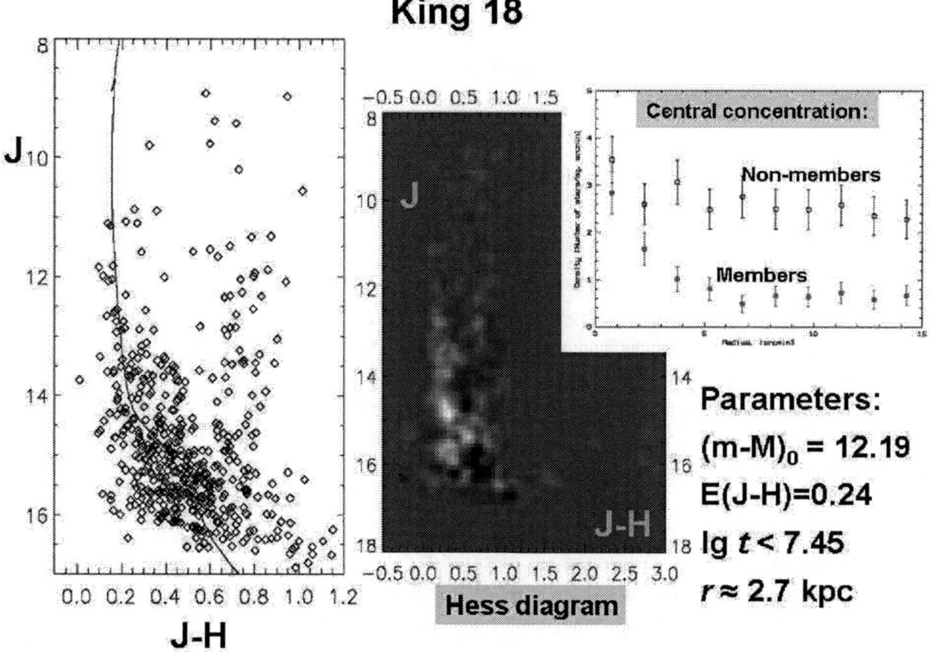

FIGURE 2. 2MASS: $J-(J-H)$ diagram for King 18 cluster; Padova isochrone [6] best-fit and basic parameters. Upper right panel: Surface density for probable members and non-members.

NEAR-INFRARED PHOTOMETRY

Our first attempt to derive basic parameters of King 18 open cluster was based on Two Micron All Sky Survey (2MASS) JHK photometry [5]. Isochrone fitting technique has been applied to stars in the cluster area which concentrate to the cluster center; they are supposed to be probable cluster members. The stars adjacent to isochrone show evident concentration to the cluster center, in contrast to the field stars. The cluster has been identified as very young, with an age $\sim$ 30 Myr, locating at the distance $\sim$ 2.7 kpc, interstellar absorption being $A_V \approx 2.3^m$ (fig. 2). King 18 seems to be loosely populated open cluster.

THE OBSERVATIONS AND DATA REDUCTION

Photometric observations of this cluster in the BVR_CI_C bands have been carried out with one-meter Zeiss reflector of the Russian Special Astrophysical Observatory on September 15 and 18, 2004, with 2068×2072 pix CCD EEV 42-40. Field of view size is 7.1 arcmin with the image scale $0.207''/pix$. Digital CCD controller provides the

TABLE 1. Log of CCD observations

	September 15	September 18
B	$2\times300s$	$1\times300s$
V	$2\times300s$	$2\times300s$
R_C	$2\times300s$	$2\times300s$
I_C	$2\times300s$	$1\times300s$

gain $2.08e^-/ADU$, and read-out noise $\sim 4\ e^-$.

Photometric study is based on several frames in BVR_CI_C system made with 300 sec exposure (Table 1). We have observed open cluster NGC 7790 as primary photometrical standard simultaneously with King 18 cluster.

To avoid the problems arising in crowded fields we performed PSF-photometry using DAOPHOT/ALLSTAR software [7] in ESO-MIDAS environment. The PSFs were build up separately for each frame using 10-15 uncrowded stars. We constructed the aperture grows curves required for determining the difference between aperture and profile-fitting magnitudes to set the local standards.

We corrected our instrumental magnitudes for the atmospheric extinction by adding mean seasonal corrections derived earlier for SAO conditions; for unit air mass they are equal to $k_B = 0.25^m, k_V = 0.15^m, k_{R_C} = 0.12^m, k_{I_C} = 0.10^m$.

Transformation to the standard photometric system was made on the base of Stetson's CCD-photometry for NGC 7790 [8]. Transformation coefficients normalized to 100 second exposure time have been calculated by ~ 50 stars with the standard technique [9]:

$$September 15$$

$$\begin{aligned} V &= V_{obs} - (0.05\pm0.02)(B-V) + (2.57\pm0.01)^m \\ (B-V) &= (1.19\pm0.01)(B-V)_{obs} + (0.10\pm0.01)^m \\ (V-R_C) &= (0.81\pm0.01)(V-R)_{obs} + (-0.04\pm0.01)^m \\ (R_C-I_C) &= (0.99\pm0.03)(R-I)_{obs} + (0.84\pm0.01)^m \end{aligned}$$

$$September 18$$

$$\begin{aligned} V &= V_{obs} - (0.08\pm0.02)(B-V) + (2.56\pm0.06)^m \\ (B-V) &= (1.20\pm0.01)(B-V)_{obs} + (0.05\pm0.01)^m \\ (V-R_C) &= (0.80\pm0.02)(V-R)_{obs} + (-0.03\pm0.01)^m \\ (R_C-I_C) &= (0.99\pm0.02)(R-I)_{obs} + (0.83\pm0.03)^m \end{aligned}$$

Relatively large error in V zero-point on the frames dated by September 18 may be due to the variations of the atmospheric extinction during this night. In the next section we will see that extinction variations affects also the color and magnitudes errors.

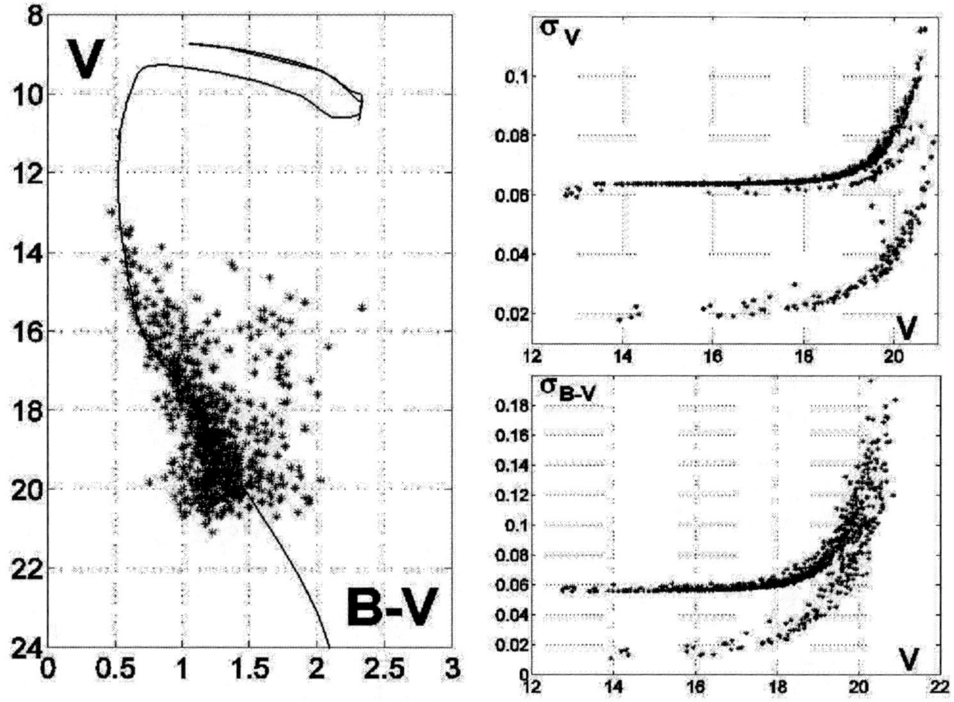

FIGURE 3. $V-(B-V)$ diagram and errors in V magnitudes and $(B-V)$ colors.

THE RESULTS

The resulting sample includes ~ 800 stars.

The apparent color-magnitude diagrams $V-(B-V), V-(V-R_C), V-(V-I_C)$ are shown in figs. 3 (left panel) and 4 with best-fit Padova isochrones [6] for solar chemical composition with overshooting.

Figs. 3 (right panel) and 5 show the errors in magnitudes and colors. Bottom sequences in these pictures refer to the frames from September 15 only, whereas upper sequences refer to combined data from September 18 and 15.

Interstellar absorption was estimated by color excess ratio taken from [10] in accordance with power-law extinction curve. True distance moduli, color excesses and interstellar absorption are listed in fig. 6. The scatter of distance moduli can be explained by atmospheric extinction variations during observations and small deviations of used filters from standard transmission curves, but optical and NIR data are in general agreement.

The average value of distance modulus from all data including NIR is $(m-M_0) \approx 12.29^m \pm 0.14^m$ and the distance equals (2.9 ± 0.2) kpc. The cluster lies near the inner boundary of the Perseus spiral arm ~ 9 kpc from the galactic center. In the future we plan continuing the photometric study of Perseus open star clusters.

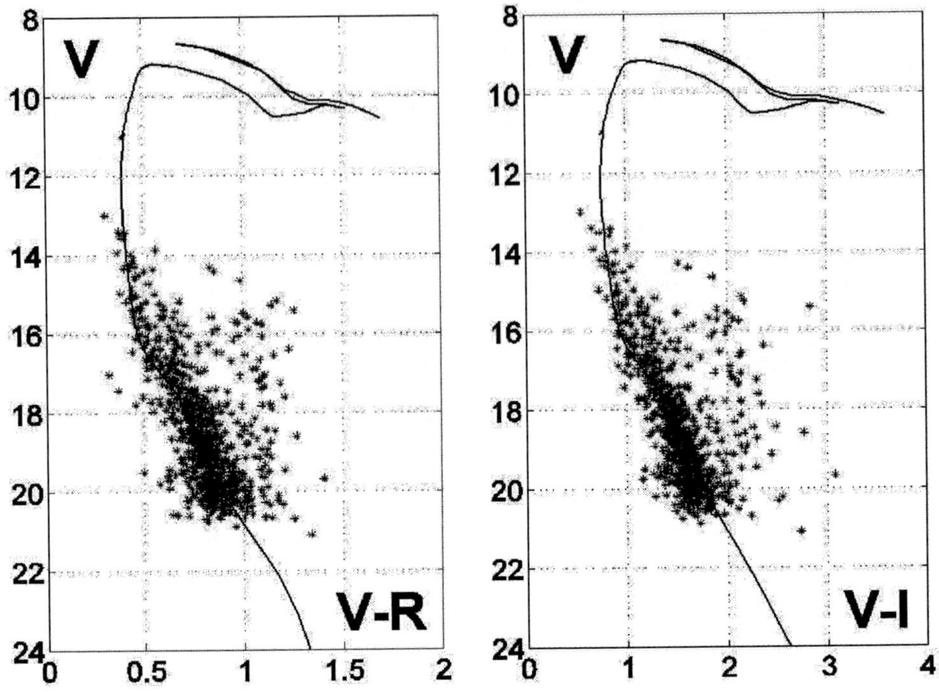

FIGURE 4. $V-(V-R_C)$ and $V-(V-I_C)$ diagrams

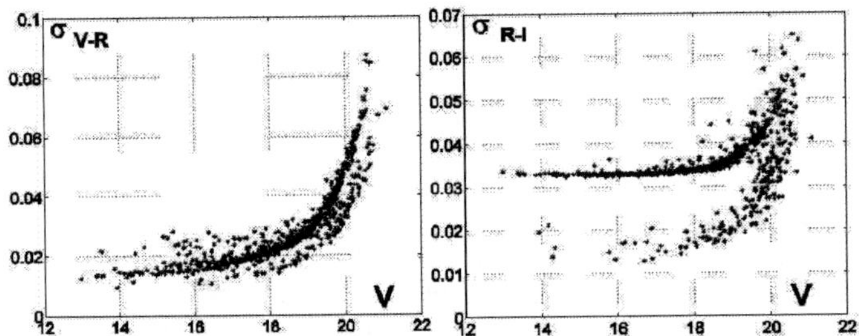

FIGURE 5. The errors in $(V-R_C)$ and $R_C-I_C)$ color indices.

ACKNOWLEDGMENTS

Authors are very grateful to our colleagues Vlasyuk V.V., Berdnikov L.N., Shatsky N.I., Kornilov V.G., Ezhkova O.V., Vozyakova O.V. and Kourilova T.S. for helpful discussions; to Russian Foundation for Basic Research (grants 04-02-16689, 05-02-16526, 05-02-16289) and to President's program for leading scientific schools support (grant NSh-5290.2006.2) for financial support.

(V, B-V)	(V, V-R_C)	(V, V-I_C)
$E(B-V) \approx 0.72^m$	$E(V-R_C) \approx 0.48^m$	$E(V-I_C) \approx 0.96^m$
$(m-M)_0 \approx 12.48^m$	$(m-M)_0 \approx 12.17^m$	$(m-M)_0 \approx 12.40^m$
$A_V \approx 2.22^m$	$A_V \approx 2.42^m$	$A_V \approx 2.20^m$

(J, J-H)	(K, J-K)
$E(J-H) \approx 0.24^m$	$E(J-K) \approx 0.38^m$
$(m-M)_0 \approx 12.19^m$	$(m-M)_0 \approx 12.23^m$
$A_V \approx 2.26^m$	$A_V \approx 2.32^m$

FIGURE 6. The parameters of King 18 cluster derived from optical and NIR data

REFERENCES

1. I. Zolotukhin, S. Koposov and E. Glushkova, Search for New Open Clusters in Huge Catalogues, *astro-ph/0601691v1* (2006)
2. E. Bica, C.M. Dutra, J. Soares and B. Barbuy, *Astron. Astrophys.* **404**, pp.223-232 (2003)
3. V.D. Ivanov, J. Borissova, P. Pessev et al., *Astron. Astrophys.* **394**, pp.L1-L4 (2002)
4. M. Kroenberger, P. Teutsch, B. Alessi et al. *Astron. Astrophys.* **447**, pp.921-928 (2006)
5. M.F. Skrutskie , R.M. Cutri , R. Stiening et al., The Two Micron All Sky Survey (2MASS), *Astron. J.* **131**, pp. 1163-1183 (2006)
6. L. Girardi, G. Bertelli, A. Bressan et al., *Astron. Astrophys.* **391**, pp. 195-212 (2002)
7. P.B. Stetson, *Publ. Astron. Soc. Pacific* **99**, pp.191-222 (1987)
8. P.B. Stetson, *Publ. Astron. Soc. Pacific* **112**, pp.925-931 (2000)
9. L.N. Berdnikov, *Perem. Zvezdy* **22**, pp. 369-400 (1986)
10. L. He, D.C.B. Whittet, D. Kilkenny, J.H. Spencer Jones, *Astrophys. J. Suppl. Ser.*, **101**, pp.335-346 (1995)

The Moscow Program of Cepheid Investigations

Rastorguev A.S.[*,†], Berdnikov L.N.[†], Samus N.N.[**,†], Gorynya N.A.[**,†], Sachkov M.E.[**], Dambis A.K.[†] and Zabolotskikh M.V.[†]

[*]*Faculty of Physics, Moscow University, 1, bld. 2 Leninskie Gory, Moscow, 119992, Russia*
[†]*Sternberg Astronomical Institute, Moscow University, 13 Universitetskii prospect, Moscow, 119991, Russia*
[**]*Institute of Astronomy, Russian Academy of Science, 48 Pyatnitskaya str., Moscow, 119017, Russia*

Abstract. The results of long-term observational program of classical cepheid investigations carried out in Moscow from 1980th, are presented. The data base includes more than 75000 UBVRI photoelectric and CCD brightness measurements for more than 650 galactic Cepheids, and approximately 10000 precise measurements of radial velocities for 165 Northern Cepheids. Classical Cepheids as stars with the most reliable distance scale play an important role in all investigations of the structure and dynamics of the galactic disk. Being observed on 100 year time interval, more than 90% of cepheids show systematic period changes. New observations are necessary to improve cepheid period - luminosity relations by taking into account the difference in their evolutionary stages and misidentification of pulsation modes .

Keywords: variable stars, distance scale, classical cepheids, galactic structure and kinematics, stellar evolution
PACS: 06.30.Bp, 97.10.Cv, 97.30.-b, 97.80.-d, 98.35.Eg, 98.62.-g

INTRODUCTION

Classical cepheids play an important role in all astronomical applications, from stellar evolution to the structure and dynamics of the Universe, being the objects with most reliable distance scale. They can be easily recognized in external galaxies and make the distance measurement procedure straightforward. Their distances, radial velocities and proper motions allow the study of the structure and kinematics of the galactic disk. Cepheid distances can be calculated from the period - luminosity relation (PL), which is usually derived both from HIPPARCOS trigonometric parallaxes and distances of young open clusters. The construction of the distance scale is shown in fig. 1. Cepheid parallaxes provide the most important link between direct and indirect distance measurement methods.

Regardless of the fact that the distance scale was widely discussed in a large number of papers, the problem is far from its final solution. The reason is in the lack of precise trigonometric parallaxes of Cepheids, because most of these stars are very far from the Sun. We need new systematic photometric observations of Cepheids to resolve their evolutionary state and improve PL-relations.

CP934, *Flows, Boundaries, Interactions*
edited by C. Dumitrache, V. Mioc, and N. A. Popescu

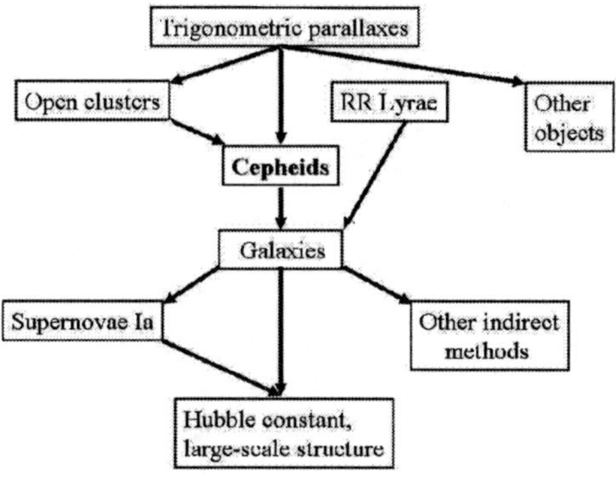

FIGURE 1. The construction of the distance scale

PHOTOMETRIC AND SPECTROSCOPIC OBSERVATIONS

Systematic photometric study of the classical cepheids started in Moscow University in 1983. The program includes more than 650 galactic stars, classified as Cepheids in the General Catalog of Variable Stars (GCVS). Multicolor photoelectric and CCD photometric observations have been carried out with the telescopes of moderate size on observatories of Russia, Uzbekistan, Ukraine, Chile, Australia and South Africa, i.e. for all Cepheids in the Northern and Southern hemispheres. The total number of UBVRI measurements of Cepheid brightness exceeds 75000. Recent photoelectric observations have been supplied by historical photographic estimates on the plates from numerous plates from Harvard and Moscow astronomical photographic collections. This has been done to create long (100 - 150 years) time series of brightness measurements and to study the period changes approximately in about 230 Cepheids. Photometrical database includes now more than 60% of all world photometric material for Cepheids [1].

Extensive spectroscopic observations of Northern Cepheids started in 1987. We have used CORAVEL-type radial velocity meter, based on correlation of spectra with the mask and constructed by A.Tokovinin. The external accuracy of radial velocity measurements approaches 0.3 km/s, and the mean accuracy of Cepheid radial velocities is approximately 0.5 km/s. Observations have been carried out with 0.6 - 1-meter telescopes in Uzbekistan, Moscow, Simeiz (Krimea, Ukraine). Since now, more than 10000 precise radial velocities of 165 Northern Cepheids have been measured, that provide good coverage of radial velocity curves. Spectral data base also includes more than 70% of all radial velocity measurements of comparable precision [2]. Radial velocities have been used for the study of cepheid kinematics, calculations of the Wesselink radii, discovery of new spectroscopic binaries among Cepheids.

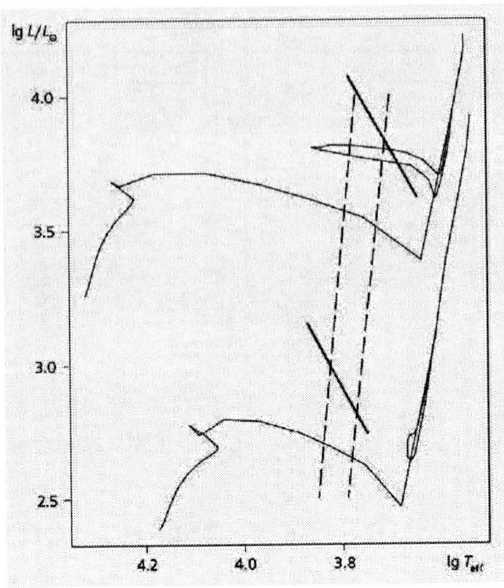

FIGURE 2. Evolutionary tracks and the instability strip

CEPHEID PERIOD CHANGES AND PERIOD - LUMINOSITY RELATION

Cepheid variables originated from B-type stars; they evolve from main sequence through supergiant area and cross the instability strip (fig. 2). The lines of constant periods are shown by inclined lines intersecting the instability strip. Because these lines are not parallel to the tracks, a Cepheid increase or decrease its pulsation period during evolution. These changes can be easily revealed on (O-C) diagrams as parabolic deviations, when analyzing long time series of brightness measurements. The time interval is long enough (more than 100 years), 90% of Cepheids to show evolutionary period variations (fig. 3).

In many cases evolutionary effects are combined with abrupt period changes.

Large statistics enabled us to find the sign and value of the evolutionary changes, compare it with theoretical estimates (fig. 4), and identify the instability strip crossing number [3]. Theoretical limits are shown by solid lines, crossing numbers are also indicated. It can be seen from the evolutionary tracks (fig. 2), that the Cepheids with the same periods may have different masses, luminosities and crossing numbers. This ambiguity, certainly, increases the scatter of the PL relations. New observations and long time series are necessary to improve classification of their evolutionary stages and multicolor PL relations.

Another source of systematic errors in the distance scale is the misidentification of the pulsation mode (fundamental - first overtone). It can be easily shown that it leads to overestimation of Cepheid's luminosity by 0.65^m, making the distance scale systematically longer. To clarify the pulsation mode of galactic Cepheids, we can use

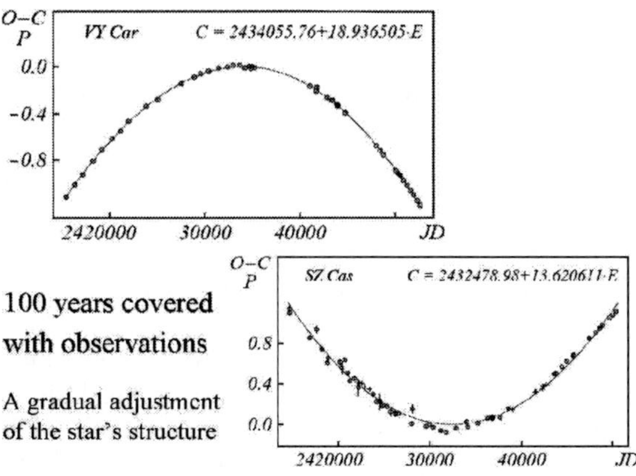

FIGURE 3. Two examples of evolutionary period changes

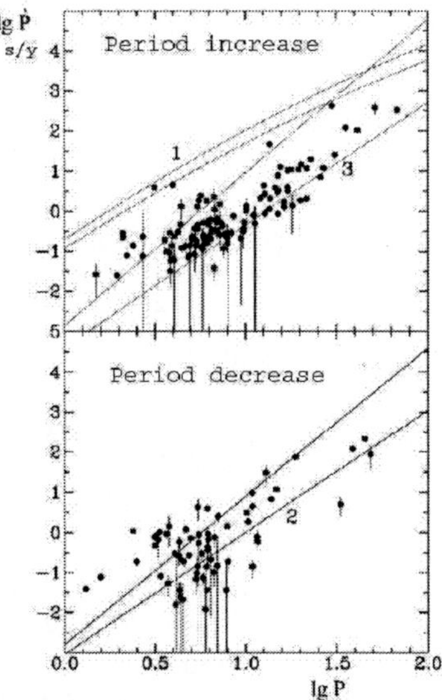

FIGURE 4. Identification of the instability strip crossing number

the Fourier-analysis of light curves or period-radius relation. Moreover, 23 double-mode Cepheids are known in our Galaxy.

The central point of Cepheid studies is the period - luminosity relation, which makes the Cepheid variables the best standard candles. We have determined its slope and zero-point from nine Cepheids - known members of young open clusters, in the $BVRR_CII_CJHK$ photometric bands. For visual band V it can be written as

$$\langle M_V \rangle_I = -1.01^m - 2.87^m \cdot \lg P_{pls} \quad (1)$$

The slopes of these PL-relations agree well with those determined for the Cepheids in the Large Magellanic Cloud. When applied to the LMC, our JHK PL-relation yields a distance modulus $(m-M)_0 \approx 18.25^m \pm 0.04^m$, in good agreement with estimations based on RR Lyrae and subdwarf stars (the "short distance scale") [4].

PULSATION RADII OF CEPHEIDS AND BINARY CEPHEIDS

Precise photometrical data and dense sets of radial velocities encouraged us to calculate Cepheid radii by straightforward Baade-Wesselink-Balona technique. Balona's modification of the technique is based on quadratic approximations of the effective temperature and bolometric correction by normal color and leads to simple regression model, which includes visible magnitude, observed color and radius changes calculated from radial velocity curve. Regression parameters also include information on the color excess and distance modulus.

We calculated the pulsation radii for 62 bright Northern Cepheids and derived the period - radius relation for fundamental-tone pulsating Cepheids [5]:

$$\lg R_F/R_\odot = (1.23 \pm 0.03) + (0.62 \pm 0.03) \lg P_{pls}, \quad (2)$$

which can be used, in particular, to separate first-overtone Cepheids. Now we started extended calculations of the radii for 165 Northern Cepheids based on new photometry and radial velocities and improved Balona technique, which takes into account new relation between radial velocity and the rate of radius change, which varies with radial velocity. Our correlation observations allowed us to discover the variations of the spectral lines width with the pulsation phase.

Our homogeneous set of radial velocity measurements has been used to study binary Cepheids and check an opinion that 70% of Cepheids are spectroscopic binary stars. We considered as possible binaries the Cepheids with noticeable (> 1-2 km/s) scatter of radial velocity curves. Our observations revealed 22-25% of Cepheids as possible binary stars, in good agreement with the same estimate for red giant stars. Radial velocity have been decomposed into three terms: first, constant velocity of the center of mass, second, Keplerian orbital motion and third, pulsation contribution; the last being expressed by the trigonometric polynomials. We discovered five new spectroscopic binaries and derived orbital parameters for approximately 20 binary Cepheids and estimated minimal masses of their companions. As a rule, orbital periods are measured by hundreds and thousands of days; the shortest periods have TX Del (133.4^d) and AU Peg (53.3^d). The classification of these stars is still uncertain; they may happen to be W Vir type halo Cepheids with much smaller radii. In this latter case it seems very intriguing that the minimal mass of AU Peg companion's ($1.2 M_\odot$) approaches the maximal mass of a white dwarf star [6].

GALACTIC STRUCTURE AND KINEMATICS

Precise distances, radial velocities, HIPPARCOS and TYCHO-2 proper motions of Cepheids make these stars the best "beacons" of the disk structure and dynamics. Together with other young populations, Cepheids outline star formation regions and spiral arms of our Galaxy. Cluster analysis of cepheid distribution reveals more than 60 complexes in 5×5 kpc field around the Sun. Their sizes vary from 0.6 to 1.4 kpc. Unfortunately, strong selection effects, which are hard to account for, make the study of spiral arms based on space distribution of young objects too difficult. Kinematical data are presumably free from these factors, and we have discovered radial and tangential periodicity of residual velocities of Cepheids, that can be considered as direct evidence for the wave nature of local spiral arms.

Combined kinematical model for the Cepheid population included pure differential rotation, deviations from rotation, "cosmic dispersion" - ellipsoidal distribution of residual velocities and motion of the sample relative to the Sun. The perturbations, due to the spiral arms considered as density waves, have been described by a simple linear theory of density waves.

The whole set of model parameters, including the pitch angle, phase angle of the Sun, and the amplitudes of radial and tangential velocity perturbations, was calculated by the maximum likelihood technique, with the allowance for the errors of observational data, model errors induced by random errors in distances, and possible systematic error of the Cepheid distance scale. This technique implements the statistical parallax approach to the Cepheid distance scale [7].

Here is the set of the calculated parameters. Rotation curve is described by angular velocity and its derivatives:
$\omega_0 = (27.5 \pm 0.5)\ km/s/kpc$, $\omega_0^{'} = (-4.5 \pm 0.15)\ km/s/kpc^2$, $\omega_0^{''} = (1.2 \pm 0.1)\ km/s/kpc^3$.
Velocity ellipsoid axes: $(\sigma_U, \sigma_V, \sigma_W) \approx (14 \pm 1, 9 \pm 0.5, 7 \pm 1)\ km/s$.
Spiral pattern parameters for two-armed Galaxy: pitch angle $i = 6^\circ \pm 0.7^\circ$.
Sun's phase angle: $\chi = -85^\circ \pm 15^\circ$.
Radial and tangential perturbation amplitudes $f_R = (7 \pm 2)\ km/s$, $f_\Theta = (-2 \pm 0.5)\ km/s$.

Combined rotation curve of our Galaxy built from radial velocities of young populations, including gas and cepheids, is shown on fig. 5 [7]. The distance to the galactic center was assumed to be $R_0 = 7.5\ kpc$. The behavior of the rotation velocity outside of 12 kpc is not very reliable.

CONCLUSION

The Cepheids remain the most important standard candles in the Local Group and in the Universe. Detailed study of evolutionary effects encourages to improve their period - luminosity relation by taking into account the differences in the evolutionary states. Some progress can be also expected from the use of the cepheid Wesselink radii and independent estimates of interstellar absorption. Continuing multicolor photometric observations and radial velocity measurements can also improve the classification of galactic Cepheids by pulsation modes.

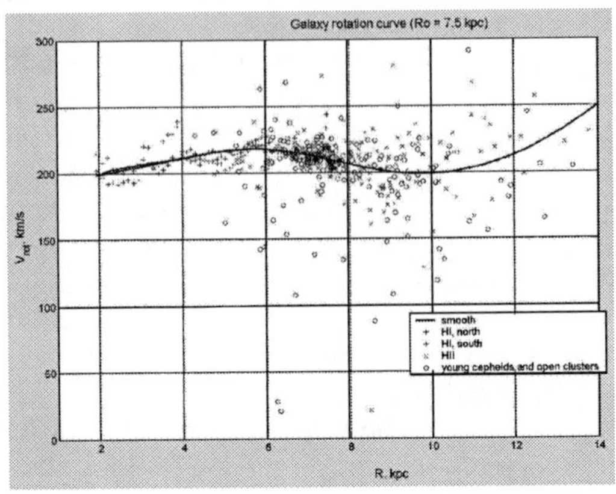

FIGURE 5. Galaxy rotation curve from radial velocities of young populations and gas

ACKNOWLEDGMENTS

We are very grateful to our colleagues Vozyakova O.V., Glushkova E.V., Antipin S.V., Pastukhova E.N. and Efremov Yu.N. for helpful discussions; to Russian Foundation for Basic Research (grants 04-02-16689, 05-02-16526, 05-02-16289) and to President's program for leading scientific schools support (grant NSh-5290.2006.2) for financial support.

REFERENCES

1. L.N. Berdnikov, The Cepheid data base, ASP Conference Series, **83**, 349-350 (2002).
2. N.A. Gorynya, N.N. Samus, M.E. Sachkov, A.S. Rastorguev, E.V. Glushkova, S.V. Antipin, Radial velocities of cepheids, VizieR On-line Data Catalog: III/229 (2002).
3. D.G. Turner, M. Abdel-Sabour Abdel-Latif, L.N. Berdnikov, Rate of period change as a diagnostic of cepheid properties, Publ. Astron. Soc. Pacific, **118**, 410-418 (2006).
4. L.N. Berdnikov, O.V. Vozyakova, A.K. Dambis, The BVRIJHK period-luminosity relations for Galactic classical Cepheids, Astronomy Letters, **2**, 838-845 (1996).
5. A.S. Rastorguev, N.A. Gorynya, N.N. Samus, Spectral binarity of cepheids, In: "Binary Stars" (ed. Massevich A.G.), 1997, Moscow: "Cosmoinform", pp.123-138.
6. M.E. Sachkov, A.S. Rastorguev, N.N. Samus, N.A. Gorynya, The radii of 62 classical Cepheids, Astronomy Letters, **24**, 377-383 (1998).
7. M.V. Zabolotskikh, A.S. Rastorguev, A.K. Dambis, Kinematic Parameters of Young Subsystems and the Galactic Rotation Curve, Astronomy Letters, **28**, 454-464 (2002).

On the Common Influence of Stark Broadening and Hyperfine Structure in Stellar Spectra : Mn II Lines

Zoran Simić*, Milan S. Dimitrijević*, Luka Č. Popović*, Miodrag Dačič*, Sylvie Sahal-Bréchot† and Andjelka Kovačević**

*Astronomical Observatory, Volgina 7, 11160 Belgrade, Serbia
†Observatorie de Paris-Meudon, 92195 Meudon, France
**Department for Astronomy, Faculty for Mathematics, Studentski Trg 16, 11000 Belgrade

Abstract. Ionized manganese lines are of interest for the analysis and modelling of stellar spectra as well as for the modelling and consideration of sub photospheric layers. Recently, a disagreement of up to 5.7 times is found between experimental and calculated Stark widths and shifts of Mn II lines. As one of possible reasons, the hfs splitting was assumed. In order to investigate the reasons for this, we performed more sophisticated calculations for two Mn II lines, by using the semiclassical perturbation theory. Moreover, we made a detailed analysis of the influence of hfs splitting on the considered experimental results. Also, the obtained results were used for the investigation of the influence of Stark broadening on Mn II spectral line profiles in DB white dwarfs. It was demonstrated the importance to take into account Stark broadening mechanism for the analysis of DB white dwarf spectra. The obtained data and conclusions are of interest for a number of problems in stellar and Solar physics, like spectrum analysis and synthesis, radiative transfer and modelling of sub photospheric layers.

Keywords: plasma; stars: lines profiles, white dwarfs
PACS: 32.70.Jz; 95.30.Dr; 97.10.Ex

INTRODUCTION

Spectral lines of the ionized manganese are present in stellar spectra and even a special class of chemically peculiar stars, the so called HgMn stars [see e.g. 1, 2] exists. For plasma conditions in hot star atmospheres, as Ap stars or white dwarfs, hydrogen is mainly ionized and Stark broadening is the main pressure broadening mechanism, influencing spectral line shapes. For example in [3, 4] is shown that in atmospheres of A stars exist conditions where Stark widths are larger than, or comparable with, the corresponding thermal Doppler widths. Stark broadening parameters, spectral line full width at half intensity maximum (FWHM) - W, and shift - d for 16 Mn II multiplets, are determined within the modified semiempirical approach [5, 6] in [7, 8]. Additionally, Stark widths for Mn II a^5D - z^5P^o multiplet, are obtained in [9] by using the semiclassical theory [10, 11], the modified semiempirical theory [5] and Griem's semiempirical theory [12]. Moreover, for Mn II a^7S - z^7P^o multiplet, width and shift, estimated on the basis of regularities and systematic trends, are given in [13].

In an experimental study [14] are determined Stark broadening widths and shifts for 11 Mn II lines, and a disagreement up to a factor of 5.7 with theoretical results of [7]was found. As possible reasons in [14]are given: (i) "It should be pointed out that the calcu-

CP934, *Flows, Boundaries, Interactions*
edited by C. Dumitrache, V. Mioc, and N. A. Popescu

lations [7] are performed using the modified semiempirical approximation which gives, generally, lower W values than the more sophisticated semiclassical approximation." (ii) "Inclusion of the helium ions, as perturbers, in the sophisticated semiclassical theory may lead to increase of the theoretical Mn II and Mn III Stark width values in helium plasma. Future calculations in this way would be helpful."

Also, the hfs splitting is discussed in Ref. [14] and the authors emphasize that due to the large Doppler broadening they had no possibility to monitor Stark broadening of particular hfs component, but only "equivalent light intensity distribution caused by Doppler broadening."

One of the aims of this contribution is to analyze disagreement of experimental [14] and theoretical results [7], but also to investigate the influence of Stark broadening mechanism on Mn II line profiles in DB white dwarf spectra and to provide new Mn II data needed for stellar plasma analyzing and modelling. Consequently, we performed more sophisticated than in [7], calculations for six Mn II lines, by using the semiclassical perturbation theory ([15, 16, 17], see also a review of updating and innovations in [19, 18]). Additionally, ionized helium impact broadening calculations were performed and a detailed analysis of the influence of hfs splitting on the experimental results of [14] was made.

CALCULATIONS WITHIN THE SEMICLASSICAL THEORY

For the calculations, the semiclassical perturbation formalism, developed and discussed in detail in [15, 16] was used. All details of the calculations and the complete results for the Stark broadening parameters for six Mn II lines will be given in [20]. As an example of results obtained, in Table 1, electron-impact broadening parameters (full width at half maximum W and shift d) for 2 Mn II lines for perturber density of $10^{17}\mathrm{cm}^{-3}$ and temperatures from 5000 to 100000 K, are shown. In order to check the possible influence of configuration interactions on disagreement between the experiment and theory, the first set of values is calculated with the included estimated maximal contribution of forbidden transitions registered in [21]. The second set of values, denoted by ($'$), is calculated taking into account only dipoly allowed transitions, as in [7, 8] [also see 5].

In Table 2, our calculations are compared with experimental results in [14]. One can see that one part of disagreement may be explained by more sophisticated, semiclassical perturbation calculations, inclusion of configuration mixing and proton-impact broadening, but the disagreement of up to a factor of 2.39 for the width is still large.

HYPERFINE STRUCTURE - HFS

In ionized manganese spectrum the hfs splitting may influence on spectral line profiles [14, 22, 23, 24]. It is stated in [14] that it is not possible to measure the hfs components, but only the whole line profile, due to large Doppler width. Here we will consider how much the hfs splitting can affect the measured [14] Doppler and Lorentz line widths. Note here, that all hfs components of a line should be with the same Lorentz and Doppler

TABLE 1. Electron-impact broadening parameters (full width at half maximum W and shift d in [Å]) for 2 Mn II lines for perturber density of $10^{17}cm^{-3}$ and temperatures from 5000 to 100000 K. The first set of values is calculated with the included estimated maximal contribution of forbidden transitions registered in [21]. The second set of values, denoted by ($'$), is calculated taking into account only dipoly allowed transitions.

Transition	T[K]	W_e[Å]	d_e[Å]	W'_e[Å]	d'_e[Å]
	5000	0.128	0.236E-03	0.141	0.924E-04
	10000	0.948E-01	-0.996E-03	0.102	-0.756E-03
a ^{7}S - z ^{7}P^o	20000	0.702E-01	-0.116E-02	0.740E-01	-0.858E-03
2594.5Å	30000	0.598E-01	-0.956E-03	0.621E-01	-0.726E-03
	50000	0.507E-01	-0.128E-02	0.516E-01	-0.924E-03
	100000	0.435E-01	-0.118E-02	0.433E-01	-0.921E-03
	5000	0.226	-0.394E-01	0.176	-0.653E-03
	10000	0.165	-0.302E-01	0.130	-0.253E-02
a ^{5}S - z ^{5}P^o	20000	0.121	-0.234E-01	0.969E-01	-0.258E-02
2950.1Å	30000	0.102	-0.193E-01	0.830E-01	-0.209E-02
	50000	0.884E-01	-0.168E-01	0.713E-01	-0.282E-02
	100000	0.800E-01	-0.137E-01	0.619E-01	-0.257E-02

TABLE 2. The comparison between experimental and theoretical Stark broadening parameters

Transition	W_m[Å]	d_m[Å]	$\frac{W_m}{W_{PD}}$	$\frac{d_m}{d_{PD}}$	$\frac{W_m}{W_e}$	$\frac{d_m}{d_e}$	$\frac{W_m}{W_e+W_i}$	$\frac{d_m}{d_e+d_i}$
3d^5(^{6}S)4s-3d^5(^{6}S)4p a ^{7}S - z ^{7}P^o 2594.5Å	0.182	-0.008	5.44	1.00	2.76	4.00	2.39	2.67
3d^5(^{6}S)4s-3d^5(^{6}S)4p a ^{5}S - z ^{5}P^o 2950.1Å	0.208	-0.026	4.28	2.32	1.80	1.18	1.60	0.96

widths, which depend on the conditions in the plasma.

First we simulated numerically the case where the Lorentzian (w_L) and Doppler (w_D) widths are the same, but taking different ratios between the widths and hfs splitting ($w_{L,D}/D_{hfs}$), assuming only two components in hfs. We calculated the sum of the two components and after that fit the total profile with Voigt function. As one can see in Fig. 1, this effect can lead to significant differences in the estimation of widths, but only in the case where $w_{L,D}/D_{hfs} < 3$. In the case where $w_{L,D}/D_{hfs} > 3$ this effect is negligible (the differences between measured widths of components and composite profiles are a few percentage). It is interesting that in the first case, the Doppler width can be overestimated in difference with the Lorentz width that is underestimated.

In order to simulate the influence of the hfs to the line profile, in this case we adopted the Doppler width given by [14] and Stark widths (w_{st}) given by [7]. First we assume that all of the profiles have the same widths (w_L and w_D), and after summation of the hfs components we obtained a composite profile (see Fig. 3). After that we fitted the composite profile with a Voigt one. From the best fit we obtain W_L and W_D. We found

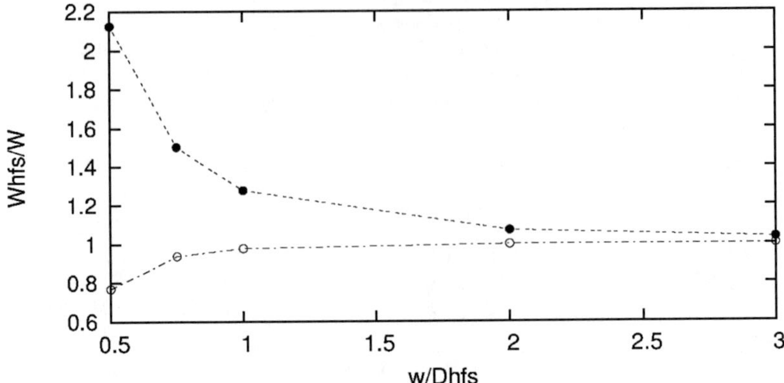

FIGURE 1. The ratio between the real widths of the *hfs* components W_{hfs} and determined ones measured (w_D is denoted with full, and W_L with open circles) from the composite profile, for different ratios of $w_{L,D}/D_{hfs}$ (where D_{hfs} is hfs splitting between two hfs components). The Lorentz and Doppler widths are taken to be the same.

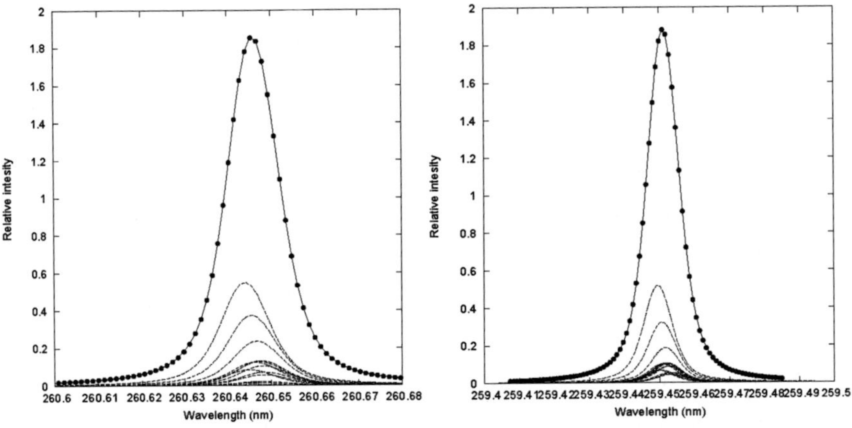

FIGURE 2. The hfs components (dashed line, below), their sum (full circles) and the best fit with the common Voigt profile (solid line) for Mn II 260.6459 nm (left) and 259.4497 nm (right). The intensity ratio of components is taken from Table 6 given in [24]

that the Doppler width is slightly larger (around 5%) and Lorentz one is slightly smaller (around 2%) than originally included in each component. Note here that the influence of hfs can affect the line shifts. In the considered cases for Mn II 260.6459 nm it is negligible, but for Mn II 259.4497 nm we found $d_{hfs} = +1.6$ pm, which is an order of magnitude of measured and calculated Stark shift for this line.

At the end, an additional test have been performed, assuming that in the case of Mn II 259.4497 nm line, each component has the Doppler width (FWHM) of 9.7 pm and Stark of 2.8 pm. Then we fixed Doppler FWHM as 5.5 pm (around 1.7 times smaller than taken in each component) and fit the composed line profile. From the fit, we obtained

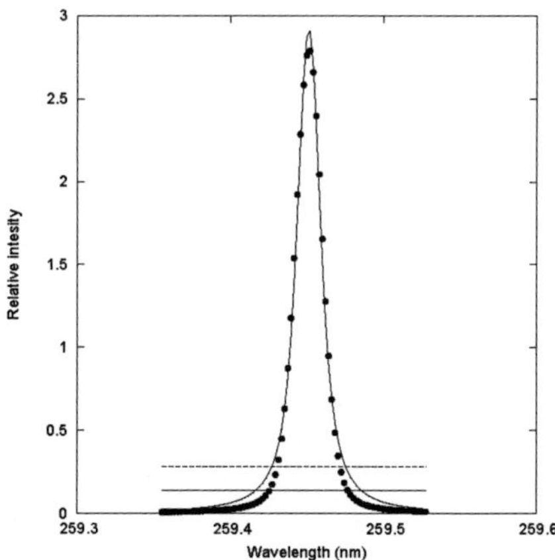

FIGURE 3. The best fit of composite line profile of Mn II 259.4497 nm where the Stark width is obtained as 2.4 times larger than was taken in each component. The horizontal lines represent the intensity at 10% (dashed line) and 5% (solid line) of maximal intensity.

that Stark width is overestimated around 2.4 times. But as can be seen in Fig. 2, there is disagreement between the best fit (solid line in the Fig. 2) and composite profile (dots in the figure) only in the wings, i.e. the part of profile with intensity smaller than 10% of maximal intensity (dashed horizontal line in the figure, solid horizontal line represents 5% of maximal intensity). Therefore, one can conclude that in the case where the line wings in experiment are not well defined/measured (e.g. a big noise in line wings), it may significantly affect the measured Stark widths.

As it can be seen from tests performed above, there is possibility that the Lorentz contribution (as well as Doppler) to the composite line profile in the case of hfs can be overestimated (underestimated). The effects of hfs on measured value of Stark width in [14] might be present. Consequently, a part of disagreement between theoretical and measured Stark broadening parameters might be due to this effect, but the question of such big differences between measured [14] and calculated [7] values of Stark width stay open.

INFLUENCE OF STARK BROADENING IN DB WHITE DWARF ATMOSPHERES

In order to investigate the importance of Stark broadening mechanism in DB white dwarf atmospheres the atmospheric models of Wickramasinghe [25], with T_{eff} = 15000 K and log g from 7 to 9, are used. Here, g is the gravitational acceleration on the stellar surface and log g=7 means that $g = 10^4$ m/s. Calculated thermal Doppler and Stark widths as a function of optical depth, for Mn II a ^{5}S - z ^{5}P^o (2950.1 Å) spectral line, are compared in Fig. 4 for DB white dwarfs. As in [25], optical depth points at the standard wavelength 5150 Åare used. As one can see, for DB white dwarf atmospheres the Stark broadening mechanism is important, especially for atmospheric layers with the optical depth larger

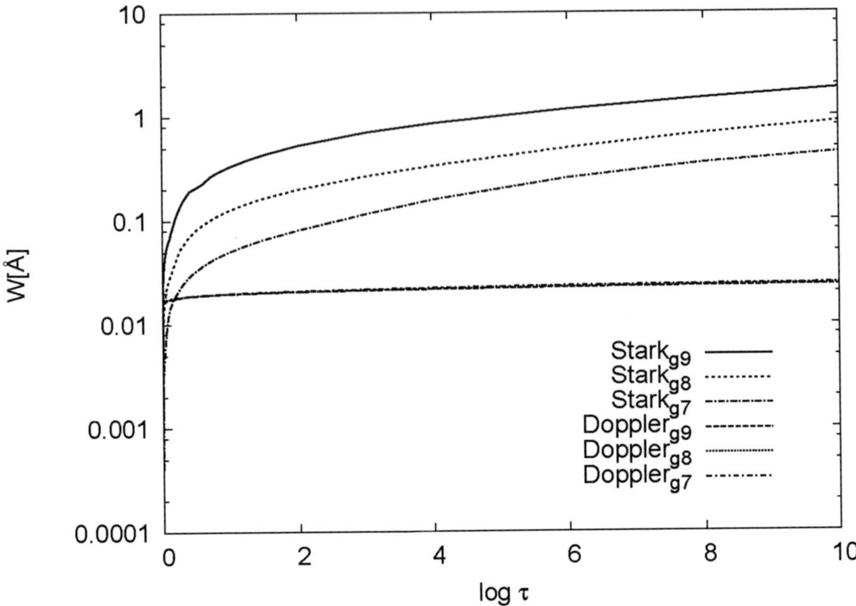

FIGURE 4. Thermal Doppler and Stark widths for Mn II spectral line a ^{5}S - z ^{5}P^o (2950.1Å) as a function of optical depth for DB white dwarf models [25]with T_{eff}=15000 K and log g from 7 to 9.

or approximatively equal to 0.1, where the Stark width is up to one or two orders of magnitude larger than the thermal Doppler width.

Consequently, in DB white dwarf atmospheres exist conditions, where Stark broadening is the principal broadening mechanism influencing line shape formation.

CONCLUSIONS

From our investigation one can conclude:

a) In DB white dwarf atmospheres the Stark broadening is important and should be taken into account.

b) The SC theory gives larger widths than MSE, and consequently the calculated values of Stark widths are in better agreement with experimental ones given in [14], but agreement between experiment and theory ($w_{exp}/w_{th} \approx 1.80 - 2.76$) is still not good.

c) The inclusion of proton-impact contribution, and configuration interaction effects could not explain the difference between experiment and theory ($w_{exp}/w_{th} \approx 1.60 -$ 2.39).

d) The hfs effects can affect measured value of Stark widths, especially if the Doppler contribution is underestimated in fitting procedure of the composite line profile. In this case the Stark width can be significantly overestimated (e.g. Mn II 259.4497 nm), especially if the wings of measured line are with big noise.

At the end, let us stress that new measurements of Stark broadening parameters for Mn II lines are needed, especially in the plasma with lower temperature in order to explain such big differences between the experimental and theoretical values.

ACKNOWLEDGMENTS

This work is a part of the projects 146001 "Influence of collisional processes on astrophysical plasma lineshapes" and 146002 "Astrophysical Spectroscopy of Extragalactic Objects" supported by the Ministry of Science of Serbia.

REFERENCES

1. G. M. Wahlgren, S. Hurbig, *A&A*, **418**, 1073 (2004).
2. K. C. Smith, M. M. Dworetsky, *A&A*, **274**, 335 (1993).
3. L. Č. Popović, M. S. Dimitrijević, *A&AS*, **120**, 373 (1996).
4. L. Č. Popović, S. Simić, M. S. Dimitrijević, N. Milovanović, *ApJS*, **135**, 109 (2001).
5. M. S. Dimitrijević, N. Konjević, *J. Quant. Spectrosc. Radiat. Transfer*, **24**, 451 (1980).
6. M. S. Dimitrijević, V. Kršljanin, *A&A*, **165**, 269 (1986).
7. L. Č. Popović, M. S. Dimitrijević, *A&AS*, **128**, 203 (1997).
8. L. Č. Popović, M. S. Dimitrijević, *Serb. Astron. J*, **156**, 173 (1997).
9. M. S. Dimitrijević, in: *Sun and Planetary System*, eds. Fricke, W., Teleki, G., D. Reidel P.C., 101 (1982).
10. W. W. Jones, S. M. Benett, H. R. Griem, Univ. of Maryland, Techn. Rep. No 1971-128, College Park, Maryland, (1971).
11. H. R. Griem, *Spectral Line Broadening by Plasmas*, McGraw-Hill, New York, (1974).
12. H. R. Griem, *Phys. Rev.*, **165**, 258 (1968).
13. I. S. Lakićević, *A&A*, **127**, 37 (1983).
14. S. Djeniže, S. Bukvić, A. Srećković, Z. Nikolić, *New Astr.*, **11**, 256 (2006).
15. S. Sahal-Bréchot, *A&A*, **1**, 91 (1969).
16. S. Sahal-Bréchot, *A&A*, **2**, 322 (1969).
17. S. Sahal-Bréchot, *A&A*, **35**, 321 (1974).
18. M. S. Dimitrijević, S. Sahal-Bréchot, *Physica Scripta*, **54**, 50 (1996).
19. M. S. Dimitrijević, *Zh. Priklad. Spektrosk.*, **63**, 810 (1996).
20. L. Č. Popović, M. S. Dimitrijević, Z. Simić, M. Dačić, S. Sahal-Bréchot, A. Kovačević, *New Astr.*, in press (2007).
21. S. Bashkin, J. O. Stoner, *Atomic Energy Levels and Grotrian diagrams*, Vol IV, Manganese, I-XXV, North Holland, Amsterdam, (1982).
22. A. J. Booth, D. E. Blackwell, *MNRAS*, **204**, 777 (1983).
23. R. A. Holt, T. J. Scholl, S. D. Rosner, *MNRAS*, **306**, 107 (1999).
24. R. J Blackwell-Whitehead, A. Toner, A. Hibbert, J. Webb, S. Ivarsson, *MNRAS*, **364**, 705 (2005).
25. D. T. Wickramasinghe, *Mem. R. Astron. Soc.*, **76**, 129 (1972).

Te I Stark Broadening Data for Stellar Plasma Analysis

Milan S. Dimitrijević*, Zoran Simić*, Andjelka Kovačević†, Miodrag Dačič* and Sylvie Sahal-Bréchot**

*Astronomical Observatory, Volgina 7, 11160 Belgrade, Serbia
†Department for Astronomy, Faculty for Mathematics, Studentski Trg 16, 11000 Belgrade
**Observatorie de Paris-Meudon, 92195 Meudon, France

Abstract. In spite of the fact that tellurium is one of the least abundant element in the Earth's litosphere, its cosmic abundance is larger than for any element with atomic number greater than 40, and its spectral lines are observed in stellar spectra. Since the significance of trace element spectral data, including Stark broadening parameters, increases with the development of space-born spectroscopy, we investigate here theoretically the influence of collisions with charged particles on spectral lines of neutral tellurium. By using the semiclassical perturbation method, Stark widths and shifts of three Te I spectral lines, of interest for modellisation, investigation and diagnostic of stellar plasma have been obtained. Results were applied for the investigation of the influence of Stark broadening mechanism in ultraviolet, optical and infrared part of the spectrum of A-type star atmospheres.The obtained results demonstrate that, in the considered case, Stark broadening is more important in optical and infrared, than in the ultraviolet part of the spectrum, and that this effect should be taken into account for the considered, A-type stellar atmosphere model.

Keywords: Stark broadening; line profiles; atomic data; stellar atmospheres.
PACS: 32.70.Jz; 95.30.Ky; 97.10Ex

INTRODUCTION

Tellurium lines are of astrophysical interest due to their presence in stellar atmospheres. For example, in Ref [1], is reported, that in Procyon photosphere spectrum one line of tellurium is identified, and used to determine the abundance of this element. Here, we will calculate within the semiclassical perturbation approach [2, 3] the Stark broadening parameters of three Te I spectral lines as a function of electron density for temperatures between 2500 K and 50000 K, particularly interesting for stellar plasma investigations. The obtained results will be used for an analysis of the influence of Stark broadening in the A type stellar atmospheres.

THEORY

Calculations have been performed within the semiclassical perturbation formalism, developed and discussed in detail in [2, 3]. This formalism, as well as the corresponding computer code, have been optimized and updated several times (see e.g. [4, 5, 6]).

Within this formalism, the full width of a neutral emitter isolated spectral line, broadened by electron impacts, can be expressed in terms of cross sections for elastic and

CP934, *Flows, Boundaries, Interactions*
edited by C. Dumitrache, V. Mioc, and N. A. Popescu

TABLE 1. This table shows electron-, and proton-impact broadening parameters for Te I, for a perturber density of 10^{16} cm^{-3} and temperatures from 2500 up to 50000 K. The quantity C (given in Å cm^{-3}), when divided by the corresponding full width at half maximum, gives an estimate for the maximum perturber density for which tabulated data may be used. The asterisk identifies cases for which the collision volume multiplied by the perturber density (the condition for validity of the impact approximation) lies between 0.1 and 0.5. For higher densities, the isolated line approximation used in the calculations breaks down. Here, T is the temperature, W$_e$ and W$_p$ denotes full line width at half maximum in Å, while d$_e$ and d$_p$ denotes line shift in Å.

TRANSITION	T(K)	W$_e$(Å)	d$_e$(Å)	W$_p$(Å)	d$_p$(Å)
	2500	0.387E-02	0.343E-02	0.106E-02	0.925E-03
Te I	5000	0.457E-02	0.399E-02	0.118E-02	0.107E-02
5p^4 ^{3}P - 6s ^{5}S^o	10000	0.548E-02	0.467E-02	0.133E-02	0.122E-02
(2372.7 Å)	20000	0.625E-02	0.531E-02	0.149E-02	0.138E-02
C=0.57E+19	30000	0.656E-02	0.547E-02	0.159E-02	0.149E-02
	50000	0.687E-02	0.533E-02	0.173E-02	0.162E-02
	2500	0.125	0.758E-01	*0.818E-01	*0.180E-01
Te I	5000	0.146	0.912E-01	*0.842E-01	*0.215E-01
6s ^{5}S^o - 7p ^{5}P	10000	0.170	0.944E-01	0.855E-01	0.251E-01
(5125.2 Å)	20000	0.196	0.894E-01	0.865E-01	0.288E-01
C=0.57E+19	30000	0.212	0.770E-01	0.871E-01	0.311E-01
	50000	0.230	0.638E-01	0.880E-01	0.341E-01
	2500	0.143	-0.698E-01	0.680E-01	-0.188E-01
Te I	5000	0.151	-0.808E-01	0.688E-01	-0.217E-01
6s ^{5}S^o - 6p ^{5}P	10000	0.170	-0.948E-01	0.697E-01	-0.248E-01
(9903.9 Å)	20000	0.194	-0.109	0.707E-01	-0.282E-01
C=0.99E+20	30000	0.209	-0.113	0.714E-01	-0.303E-01
	50000	0.226	-0.112	0.724E-01	-0.331E-01

inelastic processes as

$$W_{if} = 2\frac{\lambda_{if}^2}{2\pi c} n_e \int v f(v) dv \left(\sum_{i' \neq i} \sigma_{ii'}(v) + \sum_{f' \neq f} \sigma_{ff'}(v) + \sigma_{el}\right) \quad (1)$$

and the corresponding line shift as

$$d_{if} = \frac{\lambda_{if}^2}{2\pi c} n_e \int v f(v) dv \int_{R_3}^{R_D} 2\pi\rho d\rho \sin 2\phi_p \quad (2)$$

Here, λ_{if} is the wavelength of the line originating from the transition with the initial atomic energy level i and the final level f, c is the velocity of light, n_e is the electron density, $f(v)$ is the Maxwellian velocity distribution function for electrons, and ρ denotes the impact parameter of the incoming electron. The inelastic cross section $\sigma_{jj'}(v)$ is determined according to Chapter 3 in [2], and elastic cross section σ_{el} according to [2]. The cut-offs (needed for the calculation of inelastic and elastic cross sections and R_3),

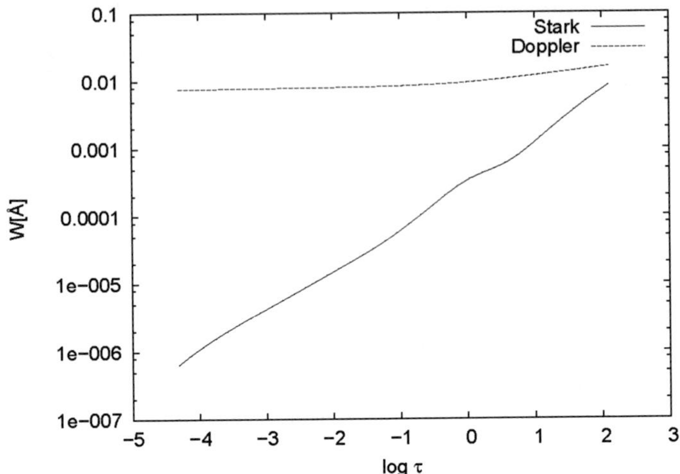

FIGURE 1. Thermal Doppler and Stark widths for Te I $5p^4\,{}^3$P - 6s ^{5}S^o (2372.7 Å) multiplet as functions of optical depth for an A type star (T_{eff} = 10000 K, log g = 4.5).

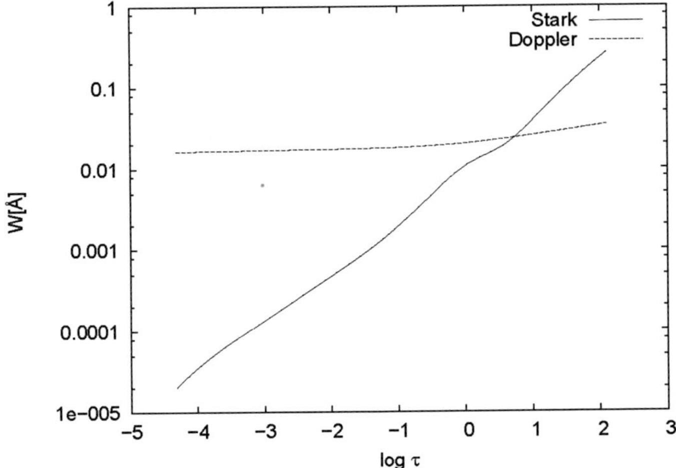

FIGURE 2. Thermal Doppler and Stark widths for Te I 6s ^{5}S^o - 7p ^{5}P (5125.2 Å) multiplet as functions of optical depth for an A type star (T_{eff} = 10000 K, log g = 4.5).

included in order to maintain for the unitarity of the S-matrix, are described in Section 1 of Chapter 3 in [2].

The formulae for the ion-impact broadening parameters are analogous to the formulae for electron-impact broadening. We note that the fact that the colliding ions would be impact in the far wings should be checked, even for stellar densities.

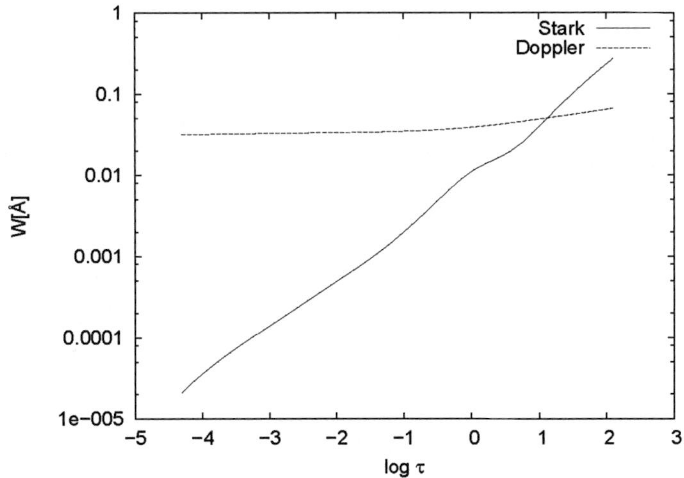

FIGURE 3. Thermal Doppler and Stark widths for Te I 6s $^5S^o$ - 6p ^5P (9903.9 Å) multiplet as functions of optical depth for an A type star (T_{eff} = 10000 K, log g = 4.5).

RESULTS AND DISCUSSION

Atomic energy levels needed for calculations have been taken from Ref. [7]. The oscillator strengths have been calculated within the Coulomb approximation ([8], and using the tables in Ref. [9]). For higher levels, the method of van Regemorter et al. has been used [10].

In Table 1, as a sample of our results, electron-, and proton-impact broadening parameters for three Te I spectral lines for a perturber density of 10^{16}cm^{-3} and temperatures from 2500 up to 50000 K, are shown. The complete results, with a detailed analysis, will be published in [11].

We used the obtained results for the investigation of the influence of Stark broadening in A type star atmospheres in different parts of the spectrum. Consequently, we took one multiplet in the ultraviolet (UV) (2372.7 Å), one in the visible (5125.2 Å), and one in the infrared (IR) (9903.9 Å) part of the spectrum.

Stark widths of 5p^4 ^{3}P - 6s $^5S^o$ (2372.7 Å) multiplet in UV have been compared in Fig. 1 with Doppler widths for a model (T_{eff}=10000 K, log g= 4.5) of A type star atmosphere [12]. Our results are presented as a function of Rosseland optical depth - log τ. The mentioned model for the stellar atmosphere has been used for two other spectral lines, for Te I 6s $^5S^o$ - 7p ^5P (5125.2 Å) multiplet in optical part of the spectrum, and for Te I 6s $^5S^o$ - 6p ^5P (9903.9 Å) multiplet in the IR (see Fig. 2. and 3.).

One can see that for the investigated case, the influence of Stark broadening mechanism is more pronounced in the optical and infrared part of the spectrum than in UV. We can see in Figs. 2 and 3 that for considered optical and IR Te I lines, exist layers in the stellar atmosphere where Stark broadening is dominant or equal to thermal Doppler broadening. For example, Stark width is larger than Doppler one for log $\tau > 0.8$ for the

5125.2 Åline. One should take into account however, that Stark broadened line profile is Lorentzian and Doppler broadened Gaussian, so that even when Stark width is smaller than Doppler one, Stark broadening might be important in the line wings.

One can conclude that, for the considered atmosphere model, Stark broadening effect should be taken into account for neutral tellurium lines, in abundance determination and other investigations of stellar plasmas.

ACKNOWLEDGMENTS

This work is a part of the projects 146001 "Influence of collisional processes on astrophysical plasma lineshapes" supported by the Ministry of Science of Serbia.

REFERENCES

1. A. V. Yuschenko, V. F. Gopka, *Astron. Astrophys. Transactions*, **10**, 307 (1996).
2. S. Sahal-Bréchot, *A&A*, **1**, 91 (1969a).
3. S. Sahal-Bréchot, *A&A*, **2**, 322 (1969b).
4. S. Sahal-Bréchot, *A&A*, **35**, 321 (1974).
5. M. S. Dimitrijević, S. Sahal-Bréchot, *A&A*, **136**, 289 (1984b).
6. M. S. Dimitrijević, *Zh. Priklad. Spektrosk.*, **63**, 810 (1996).
7. C. E. Moore, *Atomic Energy Levels III*, NSRDS-NBS 35, U.S. Govt. Printing Office, Washington D.C. (1971).
8. D. R. Bates, A. Damgaard, *Phil. Trans. Roy. Soc. London*, Ser. A, **242**, 101 (1949).
9. G. K. Oertel, L. P. Shomo, *ApJS*, **16**, 175 (1968).
10. H. van Regemorter, Hoang Binh Dy, and M. Prud'homme, *J. Phys. B*, **12**, 1073 (1979).
11. M. S. Dimitrijević, Z. Simić, A. Kovačević, M. Dačić, S. Sahal-Bréchot, *PASJ*, to be submitted (2007).
12. R. L. Kurucz, *Astrophys. J. Suppl. Series*, **40**, 1 (1979).

Influence of Collisions with Charged Particles on Solar Type Star Spectra; Investigations on Belgrade Astronomical Observatory

Milan S. Dimitrijević

Astronomical Observatory, Volgina 7, 11160 Belgrade, Serbia

Abstract. The significance of Stark broadening for Solar and stellar plasma research is discussed. Since the influence of Stark broadening within a spectral series increases with the increase of the principal quantum number of the upper level, the Stark broadening contribution may become significant for Rydberg atoms and ions where the optical electron is weakly bound to the core. Consequently, e.g. high member Balmer series lines may be used for Solar plasma diagnostic. The importance of Stark broadening for sub photospheric layers modelling is discussed also. The methods and results developed and obtained on Belgrade Astronomical Observatory and in Serbia are principally discussed .

Keywords: Stark broadening - Solar Rydberg lines - subphotospheric layers - line profiles.
PACS: 32.70.Jz; 95.30.Ky; 97.10Ex

INTRODUCTION

Development of space astronomy and computers stimulates especially the need for a large amount of atomic and spectroscopic data. Particularly large number of data is needed for example for opacity calculations and stellar modelling. For example, PHOENIX (see [1] and references therein) computer code for the stellar modelling includes a database containing data for more than hundred millions atomic/ionic and molecular transitions. The dramatic increase of accuracy and resolution is well illustrated in Fig. 1 where the χ Lupi UV spectrum obtained by IUE (International Ultraviolet Explorer) and GHRS (Goddard High Resolution Spectrograph on Hubble Space Telescope) are compared.

We will discuss here the astrophysical significance of Stark broadening data particularly for Solar and stellar plasma research and the work on this field on Belgrade Astronomical Observatory and Serbia.

ASTROPHYSICAL PLASMA CONDITIONS AND STARK BROADENING

Plasma conditions in astrophysical plasmas are much more various than in laboratory plasma sources. Consequently, broadening due to interaction between emitter and charged particles (Stark broadening) is of interest in astrophysics not only in usual but also in extreme conditions which cannot be obtained in laboratory, like in the interstellar molecular clouds or neutron star atmospheres.

CP934, *Flows, Boundaries, Interactions*
edited by C. Dumitrache, V. Mioc, and N. A. Popescu

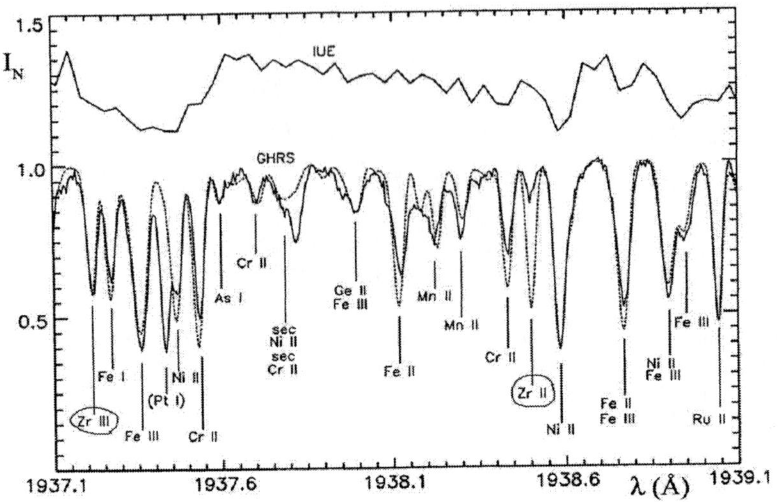

FIGURE 1. The UV spectrum of χ Lupi obtained with GHRS and with IUE satellite [2]. The full line at GHRS spectrum is observed and the dotted synthesized one

In interstellar molecular clouds, typical electron temperatures are around 30 K or smaller, and typical electron densities are 2-15 cm^{-3}. In such conditions, free electrons may be captured (recombination) by an ion in very distant orbit with principal quantum number (n) values of several hundreds and deexcite in cascade to energy levels $n-1, n-2, \ldots$ radiating in radio domain. Such distant electrons are weakly bounded with the core and may be influenced by very weak electric microfield. Consequently, Stark broadening may be significant (see e.g. [3]).

For $T_{\rm eff} > 10^4$K, hydrogen, the main constituent of stellar atmospheres is mainly ionized, and among collisional broadening mechanisms for spectral lines, the dominant is the Stark effect. This is the case for white dwarfs and hot stars of O, B and A type. Even in cooler star atmospheres as e.g. Solar one, Stark broadening may be important. For example, the influence of Stark broadening within a spectral series increases with the increase of the principal quantum number of the upper level [4-6] and consequently, Stark broadening contribution may become significant even in the Solar spectrum [7-9]. For example, high member Balmer series lines may be used as a powerful diagnostic tool in studying stellar atmospheres (see e.g. [10]).

On the other side of extreme conditions where Stark broadening is important are neutron stars. The densities of matter and electron concentrations and temperatures in atmospheres of such objects are orders of magnitude larger than in atmospheres of white dwarfs, and are typical for stellar interiors. Surface temperatures for the photospheric emission are of the order of 10^6 - 10^7 K and electron densities of the order of 10^{24} cm^{-3} [11], [12].

LINE SHAPES AND STELLAR PLASMA RESEARCH

Line shapes enter in the models of radiative envelopes by the estimation of the quantities such as absorption coefficient (κ_ν), Rosseland optical depth (τ_{Ross}) and the total opacity cross-section per atom σ_ν(op). Let us take the direction of gravity as z-direction, dealing with a stellar atmosphere. If the atmosphere is in macroscopic mechanical equilibrium and with ρ is denoted gas density, the optical depth is

$$\tau_\nu = \int_z^\infty \kappa_\nu \rho dz, \tag{1}$$

$$\kappa_\nu = N(A,i)\phi_\nu \frac{\pi e^2}{mc} f_{ij}, \tag{2}$$

where κ_ν is the absorption coefficient at a frequency ν, $N(A,i)$ is the volumic density of radiators in the state i, f_{ij} is the absorption oscillator strength, m is the electron mass and ϕ_ν spectral line profile. Let us introduce an independent variable, a mean optical depth

$$\tau_{Ross} = \int_z^\infty \kappa_{Ross} \rho dz. \tag{3}$$

For the Rosseland mean optical depth τ_{Ross}, κ_{Ross} is defined as

$$\frac{1}{\kappa_{\mathrm{Ross}}} \int_0^\infty \frac{dB_\nu}{dT} d\nu = \int_0^\infty \frac{1}{\kappa_\nu} \frac{dB_\nu}{dT} d\nu, \tag{4}$$

where

$$B_\nu(T) = \frac{2h\nu^3}{c^2}(e^{h\nu/kT} - 1)^{-1}. \tag{5}$$

Stark broadening parameters are needed as well for the determination of the chemical composition of stellar atmospheres i.e. for stellar elemental abundances determination. The method which uses synthetic and observed spectra and adjustment of atmospheric model parameters to obtain the best agreement is well developed and applied to many stars.

CALCULATION OF NEEDED STARK BROADENING PARAMETERS AND THE MODIFIED SEMIEMPIRICAL METHOD

When we need to calculate Stark broadening parameters of spectral lines, needed for solar or stellar plasma analysis one of possibilities is to use the modified semiempirical (MSE) approach [13-16], developed in Serbia for the calculation of Stark widths and shifts for non-hydrogenic ion spectral lines. From its formulation in 1980, the considered method has been applied successfully many times for different problems in astrophysics and physics.

In comparison with the full semiclassical approach [17-19] and the Griem's semiempirical approach [20] who needs practically the same set of atomic data as the more sophisticated semiclassical one, the modified semiempirical approach [13-16] needs a considerably smaller number of such data. In fact, if there are no perturbing levels strongly violating the assumed approximation, for *e.g.* the line width calculations, we need only the energy levels with $\Delta n = 0$ and $\ell_{if} = \ell_{if} \pm 1$, since all perturbing levels with $\Delta n \neq 0$, needed for a full semiclasical investigation or an investigation within the Griem's semiempirical approach [20], are lumped together and approximately estimated. Here, n is the principal and ℓ the orbital angular momentum quantum numbers of the optical electron and with i and f are denoted the initial and final state of the considered transition.

Due to the considerably smaller set of needed atomic data in comparison with the complete semiclassical [17-19] or Griem's semiempirical [20] methods, the MSE method is particularly useful for stellar spectroscopy depending on very extensive list of elements and line transitions with their atomic and line broadening parameters where it is not possible to use sophisticated theoretical approaches in all cases of interest.

The MSE method is also very useful whenever line broadening data for a large number of lines are required, and the high precision of every particular result is not so important like e.g. for opacity calculations or plasma modelling. Moreover, in the case of more complex atoms or multiply charged ions the lack of the accurate atomic data needed for more sophisticated calculations, makes that the reliability of the semiclassical results decreases. In such cases the MSE method might be very interesting as well.

SIMPLIFIED MSE FORMULA

For the astrophysical purposes, of particular interest might be the simplified semiempirical formula [15] for Stark widths of isolated, singly, and multiply charged ion lines applicable in the cases when the nearest atomic energy level ($j' = i'$ or f') where a dipolly allowed transition can occur from or to initial (i) or final (f) energy level of the considered line, is so far, that the condition $x_{\mathrm{jj'}} = E/|E_{\mathrm{j'}} - E_{\mathrm{j}}| \leq 2$ is satisfied. In such a case full width at half maximum is given by the expression (6):

$$W(\mathring{A}) = 2.2151\times 10^{-8}\frac{\lambda^2(\mathrm{cm})N(\mathrm{cm}^{-3})}{T^{1/2}(\mathrm{K})}(0.9-\frac{1.1}{Z})\sum_{\mathrm{j=i,f}}(\frac{3n_{\mathrm{j}}^{*}}{2Z})^2(n_{\mathrm{j}}^{*2}-\ell_{\mathrm{j}}^2-\ell-1). \quad (6)$$

Here, N and T are the electron density and temperature respectively, $E = 3kT/2$ is the energy of perturbing electron, $Z-1$ is the ionic charge and n the effective principal quantum number. This expression is of interest for abundance calculations, as well as for stellar atmospheres research, since the validity conditions are often satisfied for stellar plasma conditions.

Similarly, in the case of the shift

$$d(\mathring{A}) = 1.1076\times 10^{-8}\frac{\lambda^2(\mathrm{cm})N(\mathrm{cm}^{-3})}{T^{1/2}(\mathrm{K})}(0.9-\frac{1.1}{Z})\frac{9}{4Z^2}\times \quad (7)$$

$$\times \sum_{j=i,f} \frac{n_j^{*2}\varepsilon_j}{2\ell_j+1}\{(\ell_j+1)[n_j^{*2}-(\ell_j+1)^2]-\ell_j(n_j^{*2}-\ell_j^2)\}. \tag{8}$$

If all levels $\ell_{i,f} \pm 1$ exist, an additional summation may be performed in Eq. (16) to obtain

$$d(\AA) = 1.1076\times 10^{-8}\frac{\lambda^2(\mathrm{cm})N(\mathrm{cm}^{-3})}{T^{1/2}(\mathrm{K})}(0.9-\frac{1.1}{Z})\frac{9}{4Z^2}\sum_{j=i,f}\frac{n_j^{*2}\varepsilon_j}{2\ell_j+1}(n_j^{*2}-3\ell_j^2-3\ell_j-1), \tag{9}$$

where $\varepsilon = +1$ for $j = i$ and -1 for $j = f$.

SOME OTHER METHODS FOR STARK BROADENING PARAMETER DETERMINATION DEVELOPED OR USED IN BELGRADE

In a lot of cases such as e.g. complex spectra, heavy elements or transitions between highly excited energy levels, the more sophisticated quantum mechanical approach is very difficult or even practically impossible to use and, in such cases, the semiclassical approach remains the most efficient method for Stark broadening calculations.

In order to complete as much as possible Stark broadening data needed for astrophysical and laboratory plasma research and stellar opacities calculations we are making a continuous effort to provide Stark broadening data for a large set of atoms and ions. In a series of papers we have performed large scale calculations of Stark broadening parameters for a number of spectral lines of various emitters, within the semiclassical - perturbation formalism [17,18], for transitions when a sufficiently complete set of reliable atomic data exists and a good accuracy of obtained results is expected. Our semiclassical Stark broadening parameters, were used for different astrophysical problems, as e.g. for comparison with theoretical results obtained within the Stark broadening theory of solar Rydberg lines in the far infrared spectrum [21]. Also, our semiclassical results for lithium [22] have been used for a study of the non-LTE formation of Li I lines in cool stars [23] and our data for Mg I [24] for a non-LTE analysis of Mg I in the solar atmosphere [25].

When reliable data do not exist, the knowledge on regularities and systematic trends of line broadening parameters can be used for quick acquisition of new data especially when high accuracy of each particular value is not needed. This method to quickly estimate Stark broadening parameters, is mainly elaborated in Serbia (see e.g. [26-33]). The aim of such studies is to find out how regularities and systematic trends can be used to predict line widths and to critically evaluate experimental data. With the suitable use of the knowledge of regularities and systematic trends, we might use the existing experimental and theoretical values for the interpolation of new data needed in stellar spectroscopy. One must take into account however, that the validity of systematic trends and line broadening data is limited to the plasma conditions for which they are derived and extrapolations are of low accuracy.

Stark broadening research is a developed research field in Serbia, which has a critical mass of scientists and due to its often interdisciplinary significance provides a good basis for scientific collaboration. In Refs. [34-38] spectral line shapes investigations in Yugoslavia and Serbia within 1962 - 2000 period has been reviewed. It is shown that during this period 1427 (1222 by Serbian authors) bibliographic items have been published by 179 Yugoslav authors (152 from Serbia, 26 from Croatia and 1 living in France). Majority of these articles concern Stark broadening. These publications, containing also citation analysis, offer additionally a possibility to analyze the results of investigations and their applications in order to show possibilities for development of collaboration.

REFERENCES

1. E. Baron, P. H. Hauschildt, *ApJ* **495**, 370 (1998).
2. D. S. Leckrone, G. M. Wahlgren, S. G. Johansson, S. J. Adelman, in Peculiar Versus Normal Phenomena in A-Type and Related Stars, (eds. M. M. Dworetsky, F. Castelli and R. Faraggiana), *ASP Conference Series* **44**, 42, (1993).
3. A. Omont, P. Encrenaz, *A&A* **56**, 447, (1977).
4. M. S. Dimitrijević, S. Sahal-Bréchot, *J. Quant. Spectrosc. Radiat. Transfer* **31**, 301, (1984).
5. M. S. Dimitrijević, S. Sahal–Bréchot, *A&A* **136**, 289, (1984).
6. M. S. Dimitrijević, S. Sahal–Bréchot, *J. Quant. Spectrosc. Radiat. Transfer* **34**, 34, (1985).
7. I. Vince, M. S. Dimitrijević, *Publ. Obs. Astron. Belgrade* **33**, 15, (1985).
8. I. Vince, M. S. Dimitrijević, V. Kršljanin, in *Spectral Line Shapes III*, ed. F.Rostas, W.de Gruyter, Berlin, New York, p. 649, (1985).
9. I. Vince, M. S. Dimitrijević, V. Kršljanin, in *Progress in Stellar Spectral Line Formation Theory*, eds. J.Beckman and L.Crivelari, D.Reidel, Dordrecht, Boston, Lancaster, p. 373, (1985).
10. U. Feldman, G. A. Doschek, *ApJ* **212**, 913, (1977).
11. J. Madej, *A&A* **209**, 226, (1989).
12. F. Paerels, *ApJ* **476**, L47, (1997).
13. M. S. Dimitrijević, N. Konjević, *J. Quant. Spectrosc. Radiat. Transfer* **24**, 451, (1980).
14. M. S. Dimitrijević, N. Konjević, in *Spectral Line Shapes*, ed. B.Wende, W. de Gruyter, Berlin, New York, p. 211, (1981).
15. M. S. Dimitrijević, N. Konjević, *A&A* **172**, 345, (1987).
16. M. S. Dimitrijević, V. Kršljanin, *A&A* **165**, 269, (1986).
17. S. Sahal – Bréchot, *A&A* **1**, 91, (1969).
18. S. Sahal – Bréchot, *A&A* **2**, 322, (1969).
19. H. R. Griem, *Spectral Line Broadening by Plasmas*, Academic Press, New York, (1974).
20. H. R. Griem, *Phys. Rev.*, **165**, 258, (1968).
21. H. Van Regemorter, D. Hoang-Binh, *A&A* **277**, 623, (1993).
22. M. S. Dimitrijević, S. Sahal-Bréchot, *J. Quant. Spectrosc. Radiat. Transfer* **46**, 41, (1991).
23. M. Carlsson, R. J. Rutten, J. H. M. J. Bruls, N. G. Shchukina, *A&A* **288**, 860, (1994).
24. M. S. Dimitrijević, S. Sahal-Bréchot, *A&AS* **117**, 127, (1996).
25. G. Zhao, K. Butler, T. Gehren, *A&A* **333**, 219 (1998).
26. J. Purić, I. Lakićević, V. Glavonjić, *Phys. Lett.* **76a**, 128, (1980).
27. M. S. Dimitrijević, *A&A* **112**, 251, (1982).
28. W. L. Wiese, N. Konjević, *J. Quant. Spectrosc. Radiat. Transfer* **28**, 185, (1982).
29. M. S. Dimitrijević, *A&A* **145**, 439. (1985).
30. J. Purić, M. Ćuk, M. S. Dimitrijević, A. Lesage, *ApJ* **382**, 353, (1991).
31. W. L. Wiese, N. Konjević, *J. Quant. Spectrosc. Radiat. Transfer* **47**, 185, (1992).
32. J. Purić, *Zh. Prikl. Spektrosk.* **63**, 816, (1996).
33. J. Purić, M. Šćepanović, *ApJS* **521**, 490, (1999).

34. M. S. Dimitrijević, Line Shapes Investigations in Yugoslavia 1962-1985 (Bibliography and citation index), *Publ. Obs. Astron. Belgrade* **39**, (1990).
35. M. S. Dimitrijević, Line Shapes Investigations in Yugoslavia II. 1985-1989 (Bibliography and citation index), *Publ. Obs. Astron. Belgrade* **41**, (1991).
36. M. S. Dimitrijević, Line Shapes Investigations in Yugoslavia and Serbia III. 1989-1993 (Bibliography and citation index), *Publ. Obs. Astron. Belgrade* **47**, (1994).
37. M. S. Dimitrijević, Line Shapes Investigations in Yugoslavia and Serbia IV. 1993-1997 (Bibliography and citation index), *Publ. Obs. Astron. Belgrade* **58**, (1995).
38. M. S. Dimitrijević, Line Shapes Investigations in Yugoslavia and Serbia V. 1997-2000 (Bibliography and citation index), *Publ. Obs. Astron. Belgrade* **70**, (2001).

Papapetrou Energy-Momentum Complex for a Stringy Black Hole Solution

I. Radinschi[*,†] and B. Ciobanu[*,**]

[*]*Department of Physics, "Gh. Asachi" Technical University, Iasi, 700050, Romania,*
[†]*radinschi@yahoo.com*
[**]*bciobanu2003@yahoo.com*

Abstract. The aim of this paper is to evaluate the energy distribution of a stringy black hole solution with the Papapetrou energy-momentum complex. The space-time under consideration describes the dual solution known as the magnetic black hole solution. The energy distribution depends on the mass M and charge Q of the magnetic black hole.

Keywords: Papapetrou energy-momentum complex, stringy black hole solution.
PACS: 04. 20.-q, 04. 20.

INTRODUCTION

For many years, the issue of energy-momentum localization is controversial and lacks of a definite answer, and continues to be one of the most interesting and challenging problems of general relativity. We point out that this problem of defining in an acceptable manner the energy-momentum density hasn't got a generally accepted answer yet. Further, the subject of energy localization continues to be an open one since Einstein [1] has given his energy-momentum complex, which is not a tensor and not a covariant quantity. A local definition for energy-momentum is also connected to the existence of a coordinate system where the energy and momentum densities vanish at a point, and this is a permanent problem in energy-momentum localization.

The pseudotensorial definitions include the prescriptions of Einstein [1], Landau-Lifshitz [2], Papapetrou [3], Bergmann-Thomson [4] and Weinberg [5]. All these prescriptions have a drawback: they give the meaningful results if the calculations are restricted to quasi-Cartesian coordinates. On the other hand, Møller [6] proposed an energy-momentum complex which allows to calculate the energy distribution in any coordinate system, and not only in quasi-Cartesian coordinates. Chang, Nester and Chen [7] showed that the energy-momentum complexes are actually quasi-local and legitimate expressions for the energy-momentum. They concluded that there exist a direct relationship between energy-momentum complexes and quasi-local expressions because every energy-momentum complex is associated with a legitimate Hamiltonian boundary term.

The problem of energy-momentum localization applying different energy-momentum complexes was revived at the beginning of the last decade [8]-[9], and many researchers studied different space-times and obtained interesting results which demonstrate that these definitions are powerful concepts for energy-momentum localization.

In our paper we emphasize the importance of energy-momentum localization applying the Papapetrou prescription. We evaluate the energy distribution for a stringy magnetic

CP934, *Flows, Boundaries, Interactions*
edited by C. Dumitrache, V. Mioc, and N. A. Popescu

black hole solution in the Papapetrou prescription. The metric under consideration is obtained by multiplying the electric metric in the Einstein frame by a factor $e^{-2\Phi}$. The paper is organized as follows. In Sec. 2 we briefly present the magnetic black hole solution which is under study. In Sec. 3 we give a description of the Papapetrou energy-momentum complex, we compute the energy distribution for the magnetic black hole, and we also make a comparison with the values of energy obtained in the Einstein, Landau-Lifshitz, Weinberg and Møller prescriptions. Finally, in Sec. 4 we make a summary of the obtained results and some concluding remarks. Through the paper we use geometrized units ($c = 1$, $G = 1$) and follow the convention that Latin indices run from 0 to 3.

MAGNETIC BLACK HOLE SOLUTION IN STRING THEORY

In this section we present the magnetic stringy black hole solution which is known as the dual solution. String theory may be the best way to attain the holy grail of fundamental physics, which is to generate all matter and forces of nature from one basic building block. Through the years, many important studies have been made related to the string theory. About the low energy effective theory we point out that largely resembles general relativity with some new "matter" fields as the dilaton, axion etc [10]-[11]. One of its main properties is that there are two different frames in which the features of the space-time may look very different. These two frames are the Einstein frame and the string frame and they are related to each other by a conformal transformation $g^{E}_{\mu\nu} = e^{-2\Phi} g^{S}_{\mu\nu}$ which involves the massless dilaton field as the conformal factor. Kar [10] elaborated an interesting paper and obtained important results about the stringy black holes and energy conditions. He studied several black holes in two and four dimensions with regard to the Weak Energy Conditions (WEC). We think that an important issue is the study of the energy and momentum of stringy black holes.

The action for the Einstein-dilaton-Maxwell theory is given by

$$S_{EDM} = \int d^4x\sqrt{-g}e^{-2\Phi}[R + 4g_{\mu\nu}\nabla^{\mu}\Phi\nabla^{\nu}\Phi - \frac{1}{2}g^{\mu\lambda}g^{\nu\rho}F_{\mu\nu}F_{\lambda\rho}]. \tag{1}$$

Varying with respect to the metric, dilaton and Maxwell fields we get the field equations for the theory given as

$$R_{\mu\nu} = -2\nabla_{\mu}\nabla_{\nu}\Phi + 2F_{\mu\lambda}F_{\nu}{}^{\lambda}, \tag{2}$$

$$\nabla^{\nu}(e^{-2\Phi}F_{\mu\nu}) = 0, \tag{3}$$

$$4\nabla^2\Phi - 4(\nabla\Phi)^2 + R - F^2 = 0. \tag{4}$$

The metric (in the string frame) which solve the Einstein-dilaton-Maxwell field equations to yield the electric black hole is given by

$$ds^2 = A(1 + \frac{2M\sinh^2\alpha}{r})^{-2}dt^2 - \frac{1}{A}dr^2 - r^2 d\theta^2 - r^2\sin^2\theta\, d\varphi^2, \tag{5}$$

where $A = 1 - \frac{2M}{r}$.

In the string frame the dual solution known as the magnetic black hole [10] is obtained by multiplying the electric metric in the Einstein frame by a factor $e^{-2\Phi}$.

Therefore, the magnetic black hole metric is given by

$$ds^2 = \frac{A}{B}\,dt^2 - \frac{1}{AB}\,dr^2 - r^2\,d\theta^2 - r^2\sin^2\theta\,d\varphi^2, \tag{6}$$

with $B = 1 - \frac{Q^2}{Mr}$ where M is the mass and Q is the charge of the magnetic black hole.

PAPAPETROU PRESCRIPTION AND LOCALIZATION OF ENERGY

In the recent years the Papapetrou prescription yield important results which sustain the idea that the energy-momentum complexes are good tools for energy-momentum localization.

The Papapetrou energy-momentum complex [3] is given by

$$\Sigma^{ik} = \frac{1}{16\pi} N^{iklm}{}_{,lm}, \tag{7}$$

where

$$N^{iklm} = \sqrt{-g}\,(g^{ik}\eta^{lm} - g^{il}\eta^{km} + g^{lm}\eta^{ik} - g^{lk}\eta^{im}), \tag{8}$$

and

$$\eta^{ik} = diag(1,-1,-1,-1). \tag{9}$$

Σ^{00} and $\Sigma^{\alpha 0}$ are the energy and the momentum density components, respectively.

The Papapetrou energy-momentum complex satisfies the local conservation laws

$$\frac{\partial \Sigma^{ik}}{\partial x^k} = 0. \tag{10}$$

Integrating over the three-space gives the energy and momentum components

$$P^i = \iiint \Sigma^{i0}\,dx^1\,dx^2\,dx^3. \tag{11}$$

P^0 is the energy and P^α are the momentum components.

Using the Gauss theorem we obtain

$$P^i = \frac{1}{16\pi}\iint N^{i0\alpha\beta}{}_{,\beta}\,n_\alpha\,dS, \tag{12}$$

where $n_\alpha = \left(\frac{x}{r}, \frac{y}{r}, \frac{z}{r}\right)$ are the components of a normal vector over an infinitesimal surface element.

For performing the calculations using the Papapetrou energy-momentum complex we transform the metric given by (6) to quasi-Cartesian coordinates t,x,y,z according to $x = r\sin\theta\cos\varphi$, $y = r\sin\theta\sin\varphi$ and $z = r\cos\theta$. We obtain

$$ds^2 = \frac{A}{B}dt^2 - (dx^2 + dy^2 + dz^2) - \frac{\frac{1}{AB} - 1}{r^2}(xdx + ydy + zdz)^2. \tag{13}$$

In order to evaluate the energy distribution of the magnetic black hole solution, we use (7), (8), (13) and applying the Gauss theorem (12) we obtain that the expression for the energy is given by

$$E_P = \frac{r}{8}\frac{\frac{4Q^2}{Mr} - \frac{24Q^2}{r^2} + \frac{32Q^2M}{r^3} + \frac{8M}{r} - \frac{16M^2}{r^2} + \frac{16M^3}{r^3} - \frac{16Q^2M^2}{r^4}}{\left(1 - \frac{2M}{r}\right)^2\left(1 - \frac{Q^2}{Mr}\right)} \tag{14}$$

The energy distribution depends on the mass M and on the charge Q of the magnetic black hole.

In the limit $Q \to 0$, the Schwarzschild space-time, we obtain for the energy distribution the expression

$$E_P = \frac{M(r^2 - 2Mr + 2M^2)}{(r - 2M)^2}. \tag{15}$$

This expression is the same as obtained by Virbhadra [9] in the case of the Schwarzschild black hole when the calculations are done in Shwarzschild Cartesian coordinates.

In the limit $r \to \infty$, from (15) we obtain for the energy distribution of the magnetic black hole

$$E_P = M. \tag{16}$$

This is the same result as in the case of the Schwarzschild black hole when the calculations are carried out in Kerr-Schild Cartesian coordinates.

In some previous works we compute the energy distribution for the magnetic black hole in the Einstein [12], Landau-Lifshitz, Weinberg [13] and Møller [14] prescriptions. We make a comparison with the values of the energy distribution obtained in the Einstein, Landau-Lifshitz, Weinberg and Møller prescriptions in the limit $Q \to 0$. We obtain the connections

$$E_E = E_M = M \tag{17}$$

and

$$\begin{aligned} E_W &= E_{LL} = \frac{r}{(r-2M)}E_E = \frac{r(r-2M)}{(r^2 - 2Mr + 2M^2)}E_P, \\ E_P &= \frac{(r^2 - 2Mr + 2M^2)}{(r-2M)^2}E_E. \end{aligned} \tag{18}$$

For the magnetic black hole solution given by (6) the aforementioned energy-momentum complexes of Einstein, Landau-Lifshitz, Papapetrou, Weinberg and Møller do not yield the same expression for the energy distribution. Only the Weinberg and Landau-Lifshitz prescriptions allow to obtain the same result when the calculations are done in Schwarzschild Cartesian coordinates.

DISCUSSION

The different attempts made to develop a generally accepted expression for the energy-momentum density yield a plethora of energy-momentum definitions. The results obtained by several authors [8]-[9] demonstrated that the energy-momentum complexes are powerful tools for evaluating the energy and momentum in a given geometry. Chang, Nester and Chen [7] showed that the energy-momentum complexes are actually quasi-local and legitimate expressions for the energy-momentum. Furthermore, important studies have been done about the new idea of quasi-local approach for energy-momentum complexes and a large class of new pseudotensors connected to the positivity in small regions have been studied and constructed. In this light, the quasi-local quantities are associated with a closed 2-surface. The Hamiltonian boundary term determines the quasi-local quantities for finite regions and the special quasi-local energy-momentum boundary term expressions correspond each of them to a physically distinct and geometrically clear boundary condition.

In this paper we evaluate the energy distribution for a magnetic stringy black hole solution applying the Papapetrou prescription. The energy distribution depends on the mass M and charge Q of the magnetic black hole. In the limit $Q \rightarrow 0$ the result is the same as obtained by Virbhadra for the Schwarzschild space-time [9]. We compare our result with those obtained in the Einstein, Landau and Lifshitz, Weinberg and Møller prescriptions and investigate the connections between the expressions for energy obtained with the aforementioned energy-momentum complexes. These prescriptions do not provide a similar result for the energy distribution of the magnetic stringy black hole.

We conclude that we obtained different results applying the definitions of Einstein, Landau-Lifshitz, Papapetrou, Weinberg and Møller because these energy-momentum complexes are pseudotensors and are non-covariant, coordinate dependent expressions, and this agrees with the equivalence principle which states that gravity cannot be detected at a point.

ACKNOWLEDGMENTS

Work for I. Radinschi and B. Ciobanu was supported by Grant MEdC, 122, 45/2007.

REFERENCES

1. A. Einstein, Preuss. *Akad. Wiss. Berlin* **47**, 778-786 (1915); Addendum-ibid. **47**, 799-801 (1915); A. Trautman, in *Gravitation: an Introduction to Current Research*, L. Witten, Wiley, New York, 1962,

pp. 169-198.

2. L. D. Landau, and E. M. Lifshitz, *The Classical Theory of Fields*, Pergamon Press, Oxford, 1987, pp. 280-285.
3. A. Papapetrou, *Proc. R. Irish. Acad.* **A52**, 11-23 (1948.
4. P. G. Bergmann and R. Thompson, *Phys. Rev.* **89**, 400-407 (1953).
5. S. Weinberg, *Gravitation and Cosmology: Principles and Applications of General Theory of Relativity*, John Wiley and Sons, Inc., New York, 1972, pp. 165-171.
6. C. Møller, *Ann. Phys. (NY)* **4**, 347-371 (1958).
7. Chia-Chen Chang, J. M. Nester, and Chiang-Mei Chen, *Phys. Rev. Lett.* **83**, 1897-1901 (1999).
8. J. M. Aguirregabiria, A. Chamorro, and K. S. Virbhadra, *Gen. Rel. Grav.* **28**, 1393-1400 (1996); S. S. Xulu, *Int. J. Mod. Phys.* **D7**, 773-777 (1998); E. C. Vagenas, *Int. J. Mod. Phys.* **A18**, 5949-5963 (2003); Th. Grammenos, *Mod. Phys. Lett.* **A20** 1741-1751(2005); M. Sharif, and Tasnim Fatima, *Int. J. Mod. Phys.* **A20**, 4309-4330 (2005); R. Gad, *Mod. Phys. Lett.* **A19**, 1847-1854 (2004); I. Radinschi, *Mod. Phys. Lett.* **A15**, Nos. 11&12, 803-807 (2000); I. Radinschi, *Mod. Phys. Lett.* **A16(10)**, 673-678 (2001); I-Ching Yang, and I. Radinschi, *Chin. J. Phys.* **42**, 40-44 (2004); I. Radinschi, and Th. Grammenos, *Int. J. Mod. Phys.* **A21**, 2853-2861 (2006).
9. K. S. Virbhadra, *Phys. Rev.* **D60**, 104041 (1999).
10. S. Kar, *Phys. Rev.* **D55**, 4872-4879 (1997).
11. M. S. Green, J. H. Schwarz, and E. Witten, *Superstring Theory*, Cambridge University Press, 1987, pp. 140-147.
12. I. Radinschi, *Rom. J. Phys.*, Vol. **50**, Number 1-2, 57-61 (2005).
13. I. Radinschi, and B. Ciobanu, gr-qc/0608029
14. I. Radinschi, and I-Ching Yang, *New Developments in String Theory Research*, ed. Susan A. Grece, Nova Science Publishers, Inc New York, U.S.A., 2006, pp. 289-305.

Massive Star Winds and Trajectories around Black Holes

Adrian Sabin Popescu

Astronomical Institute of Romanian Academy,
Str. Cutitul de Argint 5, RO-040557 Bucharest, Romania

Abstract. At the conference EFYRA [1] we presented the reconstructed topology of the magnetic flux lines for the solar wind and for prominences, solution of the Lorentz equations of motion for the local Minkowski spacetime of a global Dimension Embedded in Unified Symmetry (DEUS) topology of the Universe. The second solution of same equations of motion will give us the massive stars wind topology, or the followed trajectories in the proximity of a black hole distorted spacetime, differentiated after the values of the integration constants.

Keywords: topology of spacetime, massive star winds, trajectories around black holes, collisions
PACS: 02.40.Ky, 97.10.Me, 97.60.Lf, 98.10.+z

INTRODUCTION

In [1] we obtained the trajectories followed by particles in the solar atmosphere and the heliosphere from the Lorentz equations of motion for the local Minkowski spacetime of a global Dimension Embedded in Unified Symmetry (DEUS) topology of the Universe. These trajectories were described, in spherical coordinates, by:

$$\begin{cases} r(t_{FRW}) = c_2 \exp\left\{-c_1 \sinh\left[\arctan\left(\dfrac{1}{\tan t_{FRW}}\right)\right]\right\} \\ \zeta(t_{FRW}) = -\arctan\left(\dfrac{1}{\tan t_{FRW}}\right) \\ \xi(t_{FRW}) = c_4 - c_3 \sinh\left[\arctan\left(\dfrac{1}{\tan t_{FRW}}\right)\right] . \end{cases} \tag{1}$$

Putting (1) in Cartesian coordinates we get (far from the source) the well known polar region's plume-like geometry and a Parker-like spiral in the equatorial region and, for different c_i values, (close to the source) prominence-like loops. The c_2 coefficient scales with the distance from the Sun, where with the generic term "Sun" we denote stellar objects with $M < 4M_{\odot}$.

CP934, *Flows, Boundaries, Interactions*
edited by C. Dumitrache, V. Mioc, and N. A. Popescu

TRAJECTORIES AROUND MASSIVE OBJECTS

The second set of solutions for the same Lorentz equations of motion:

$$\begin{cases} r(t_{FRW}) = c_2 - c_1 \sinh\left[\arctan\left(\frac{1}{\tan t_{FRW}}\right)\right] \\ \zeta(t_{FRW}) = -\arctan\left(\frac{1}{\tan t_{FRW}}\right) \\ \xi(t_{FRW}) = c_4 - c_3 \sinh\left[\arctan\left(\frac{1}{\tan t_{FRW}}\right)\right], \end{cases} \tag{2}$$

applies to the trajectories of particles around massive stars (stellar winds), for $c_1 \neq c_3$ and $c_2 \neq c_4$, or around DEUS object black holes, for $c_1 = c_3$ and $c_2 = c_4$. Writting (2) in Cartesian coordinates and plotting, we obtain the Figures 1 - 4.

In Figure 1 we illustrate the trajectories around a massive star. We can observe that, rotating the representation around the symmetry axis, the resulting geometry is similar to the bipolar ejecta of a massive star.

When $c_2 = c_4$ and constant, $c_1 = c_3$ will scale with the distance from the gravito-magnetic field's source (see Figures 2 - 4, where the distance decreases from Figure 2 to Figure 4). This particular case is special because here we are not dealing only with the geometry of the local Minkowski spacetime distorted by a black hole, but also, taking into account the self-similarity of the DEUS objects, with the trajectories followed by objects of a Friedmann-Robertson-Walker (FRW) global geometry (into a FRW Universe bubble) between Big Bang and Big Crunch. The local geometry of the FRW spacetime is Minkowski. The Big Bang/Big Crunch materializes in the central (x, y, z)=(0,0,0) point.

For a local Minkowski spacetime we observe that, going closer to the DEUS black hole, its event horizon becomes visible due to the particles "falling" on it (Figure 4 central sphere), any internal geometry beyond that being invisible to the external Minkowski observer. The external spacetime of an approaching observer becomes more and more distorted, finally (at the black hole event horizon) confounding itself with the horizon.

COLLISIONS

If we intend to study the way in which a collision between a black hole and a star of solar type occurs we should take the (2) equation (with the black hole c_i coefficients) and the (1) equation (where we note the c solar wind coefficients as d: $c_i \rightarrow d_i$). Then, we transform from spherical to Cartesian coordinates, where (x_1, y_1, z_1) are the black hole's and (x_2, y_2, z_2) the solar type star's coordinates, and make the sums $x = x_1 + x_2$, $y = y_1 + y_2$ and $z = z_1 + z_2$. The result is represented in Figure 5.

In the same way we can proceed if we intend to simulate the interaction between a black hole and a massive star, where for the massive star we write the c_i coefficients as d_i (see Figure 6).

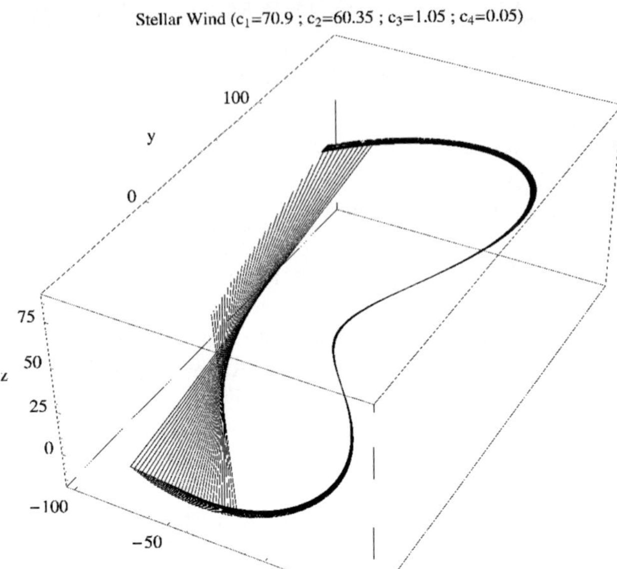

FIGURE 1.

As a consequence of the assumed DEUS symmetry of the Universe, the interaction between two black holes looks like Figure 2, where the Universe forms from the evaporation of a five-dimensional self-similar DEUS object (level one), level containing as embeddings level two DEUS objects (or, in the four-dimensional FRW spacetime, black holes).

In Figure 5 we observe that the solar type star comes from infinity, entering into the accretion disk of the black hole and, after hitting the singularity, is expelled in a jet-like feature.

The second observation that implies both 5 and 6 figures is that, considering the natural increase in mass from a solar type star to a massive star, the black hole becomes more and more affected by the gravito-magnetic field of the other body.

REFERENCES

1. A.S. Popescu, "Gravito-magnetic Heliospheric Surfaces", *AIP Conference Proceedings Series: Fifty Years of Romanian Astrophysics*, Ed. C. Dumitrache et al., ISBN 978-0-7354-0400-7 (2007).

Spacetime Topology with a DEUS Object Immersion (c_1=c_3=60.9 ; c_2=c_4=10.35)

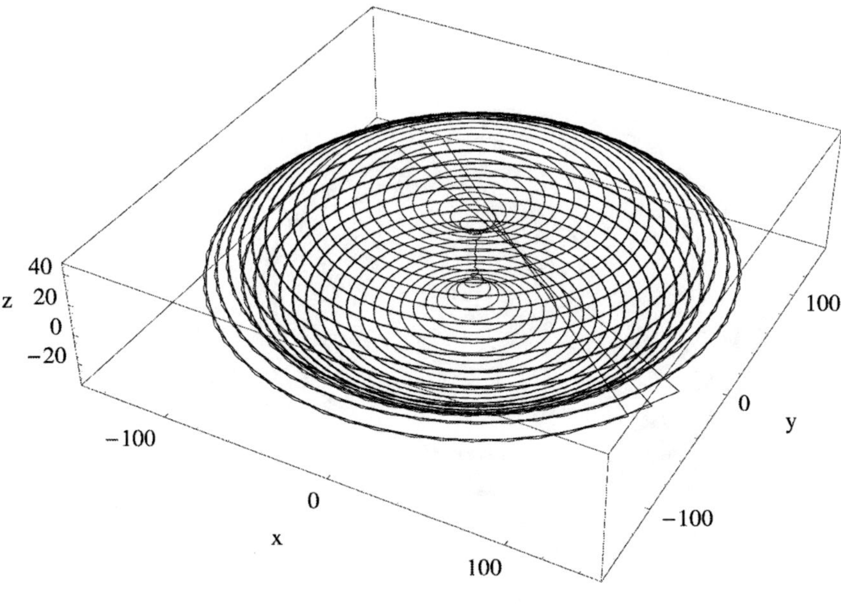

FIGURE 2.

Spacetime Topology with a DEUS Object Immersion (c_1=c_3=70.9 ; c_2=c_4=60.35)

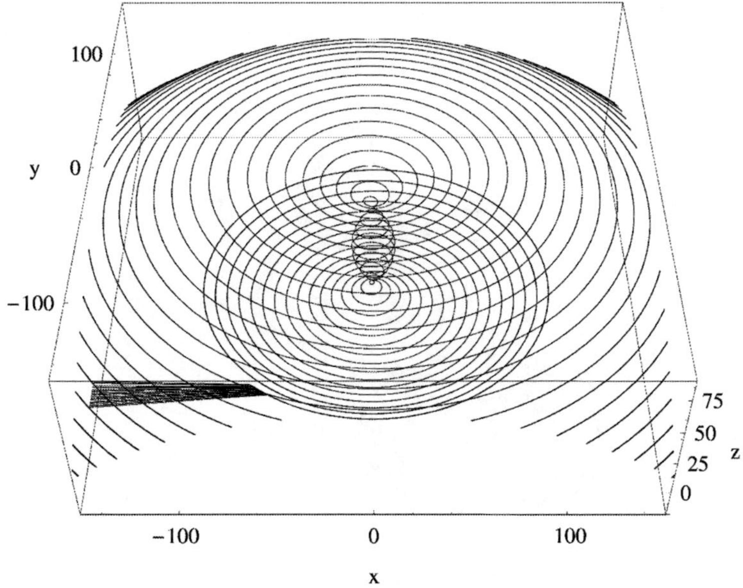

FIGURE 3.

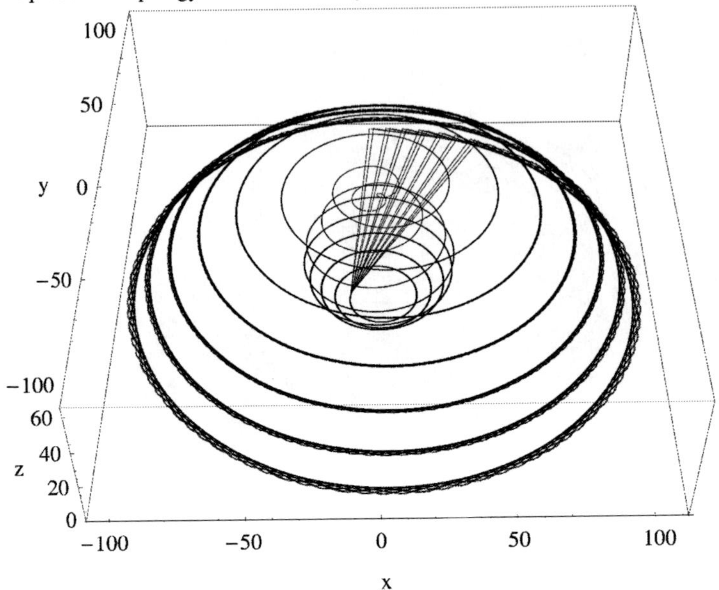

FIGURE 4.

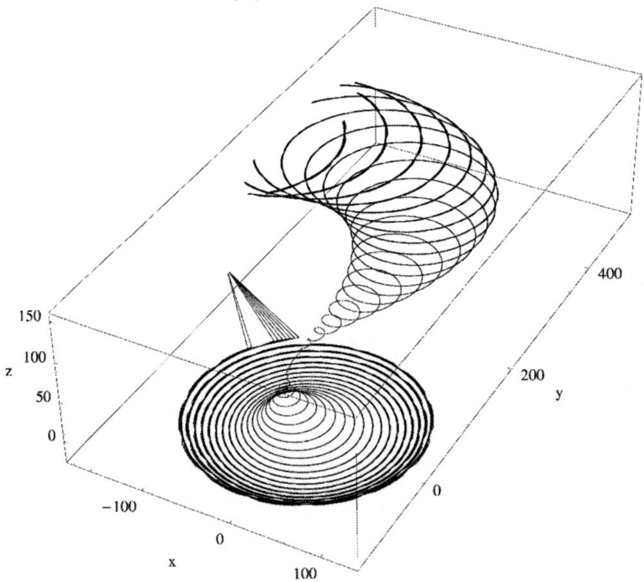

FIGURE 5.

Black Hole – Massive Star Interaction ($c_1=c_3=60.9$; $c_2=c_4=10.35$; $d_1=70.9$; $d_2=60.35$; $d_3=1.05$; $d_4=0.05$)

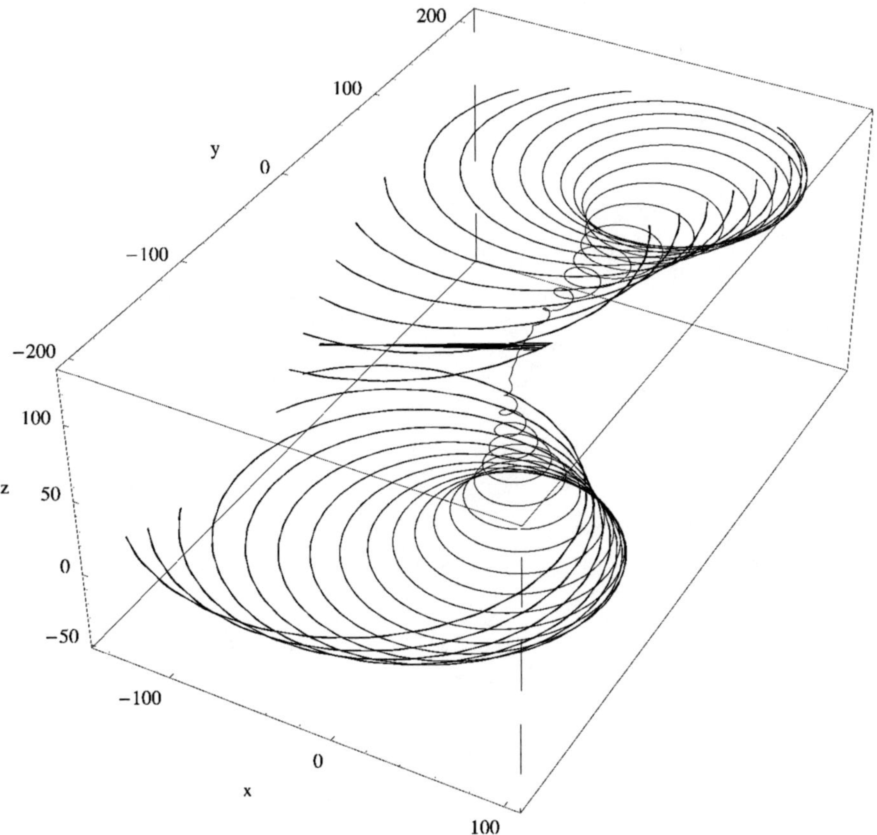

FIGURE 6.

Author Index

A

B

C

D

F

G

J

K

L

M

N

O

P

R

S

T

V

Z